Ohad Nachtomy • Reed Winegar
Editors

Infinity in Early Modern Philosophy

Springer

Editors
Ohad Nachtomy
Philosophy Department
Bar-Ilan University
Ramat Gan, Israel

Reed Winegar
Philosophy Department
Fordham University
Bronx, NY, USA

ISSN 1879-8578 ISSN 2352-2585 (electronic)
The New Synthese Historical Library
ISBN 978-3-030-06874-5 ISBN 978-3-319-94556-9 (eBook)
https://doi.org/10.1007/978-3-319-94556-9

Printed on acid-free paper

This Springer imprint is published by the registered company Springer Nature Switzerland AG
The registered company address is: Gewerbestrasse 11, 6330 Cham, Switzerland

Acknowledgments

This book was published with the support of the Israel Science Foundation (grants # 302/16 and 47/18).

Ohad Nachtomy wishes to acknowledge the assistance of Barnaby Hutchins and Noam Hoffer. Reed Winegar wishes to acknowledge the assistance of Nicholas Sooy and Christopher Meyers, the support of a Fordham University Faculty Fellowship, and the support of a VolkswagenStiftung/Mellon Postdoctoral Fellowship in the Humanities.

We are both extremely grateful to Liat Lavi for overseeing the whole project and putting everything in place.

Contents

Contributors

Igor Agostini Department of Humanistic Studies, University of Salento, Lecce, Italy

Richard T. W. Arthur Philosophy Department, McMaster University, Hamilton, Canada

Geoffrey Gorham Philosophy Department, Macalester College, Saint Paul, MN, USA

Paul Guyer Philosophy Department, Brown University, Providence, RI, USA

Mogens Lærke CNRS, IHRIM (UMR 5317), ENS de Lyon, Lyon, France

Yitzhak Y. Melamed Philosophy Department, Johns Hopkins University, Baltimore, MD, USA

Ohad Nachtomy Philosophy Department, Bar-Ilan University, Ramat Gan, Israel

Sanja Särman Philosophy Department, Hong Kong University, Hong Kong, China

Anat Schechtman Philosophy Department, University of Wisconsin-Madison, Madison, WI, USA

Noa Shein Philosophy Department, Ben-Gurion University of the Negev, Beer-Sheva, Israel

Reed Winegar Philosophy Department, Fordham University, Bronx, NY, USA

Chapter 1
Introduction: Infinity in Early Modern Philosophy

Ohad Nachtomy and Reed Winegar

In his *Pensées*, Blaise Pascal gives vivid voice to both the wonder and anxiety that many early modern thinkers felt towards infinity. Contemplating our place between the infinite expanse of space and the infinite divisibility of matter, Pascal writes:

> What is a man in the infinite?
>
> But to show him another prodigy equally astonishing, let him examine the most delicate things he knows. Let a mite be given him, with its minute body and parts incomparably more minute, limbs with their joints, veins in the limbs, blood in the veins, humours in the blood, drops in the humours, vapours in the drops. Dividing these last things again, let him exhaust his powers and his conceptions, and let the last object at which he can arrive be now that of our discourse. Perhaps he will think that here is the smallest point in nature. I will let him see therein a new abyss. I will paint for him not only the visible universe, but also everything he is capable of conceiving of nature's immensity in the womb of this imperceptible atom. Let him see therein an infinity of worlds, each of which has its firmament, its planets, its earth, in the same proportion as in the visible world; in this earth of animals, and ultimately of mites, in which he will find again all that the first had, finding still in these others the same thing without end and without cessation. Let him lose himself in wonders as amazing in their littleness as the others in their vastness. For who will not be astounded at the fact that our body, which a little while ago was imperceptible in the universe, itself imperceptible in the bosom of the whole, is now a colossus, a world, or rather a whole, in respect of the final smallness which we cannot reach? He who regards himself in this light will be afraid of himself, and observing himself suspended in the mass given him by nature between those two abysses of the Infinite and Nothing, of which he is equally removed, will tremble at the sight of these marvels; and I think that, as his curiosity changes into admiration, he will be more disposed to contemplate them in silence than to examine them with presumption.

O. Nachtomy (✉)
Philosophy Department, Bar-Ilan University, Ramat Gan, Israel
e-mail: ohadnachtomy@mac.com

R. Winegar
Philosophy Department, Fordham University, Bronx, NY, USA
e-mail: bwinegar@fordham.edu

O. Nachtomy, R. Winegar (eds.), *Infinity in Early Modern Philosophy*, The New Synthese Historical Library 76,
https://doi.org/10.1007/978-3-319-94556-9_1

> For, finally, what is man in nature? A nothing in comparison with the infinite, everything in comparison with nothingness, a mid-point between nothing and everything, infinitely removed from the two extremes. The end of things and their origin (*principes*) are invincibly hidden for him in an impenetrable secret, equally incapable of understanding the nothing from which he was drawn as the infinite in which he is surrounded.[1]

This passage, which so poignantly describes the human being's finitude, also illustrates how major scientific and mathematical developments helped spark the early modern period's preoccupation with infinity. Early modern developments in cosmology and physics, such as the introduction of a heliocentric solar system and the rise of the new mechanical science, were intimately tied to questions regarding the infinite size and infinite divisibility of space and matter. Similarly, new discoveries made through the invention of the microscope (such as smaller and smaller *animalcula*) and the development of the calculus by Leibniz and Newton raised further questions about infinite division and the infinitely small. Such questions about infinity were widely and passionately discussed by the learned public of the day. For example, questions about the infinity of space and time featured prominently in the famous correspondence between Leibniz and Newton's champion Samuel Clarke. And an eighteenth-century debate in the Berlin Academy regarding infinite division and monads grew so intense that, according to the mathematician Leonhard Euler, it spilled over into courtly society and "forced its way into company of every description, that of the guard-room not excepted" (Euler 1833: 39).

However, interest in infinity was not limited to the natural world. Rather, one defining feature of early modern philosophy is the complex interaction between the new science and traditional philosophical questions in metaphysics and rational theology. And early modern philosophers were keenly aware of infinity's theological importance. Philosophical and theological debates were, to a large extent, debates over the existence, nature, and consequences of an infinite being. Thus, for thinkers like Descartes, Leibniz, Malebranche, and Spinoza, questions about the nature of infinity were crucial for properly conceiving of God's status as the infinite being. Indeed, our contemporary distinction between science and theology is somewhat anachronistic. Under the heading of "natural philosophy," early modern thinkers normally addressed scientific and theological issues. Consequently, philosophical debates about infinity in scientific and theological contexts significantly influenced and shaped one another. For example, Descartes maintains that only God is infinite, stating "[w]e pass from knowledge of God to knowledge of his creatures by remembering that he is infinite and we are finite".[2] Descartes's commitment to God's status as the only infinite being leads him to claim that space is indefinitely, rather than infinitely, large. And he applies the same point to the divisibility of matter, which he

---

[1] The translation is ours. We have consulted the translations by Levi in Pascal (1995) and Strickland in Leibniz (2003). For the original French text, see, Pascal, Blaise. "Pensèes." In *Œuvres Complètes*, edited by Louis Lafuma. Paris: Éditions du Seuil, 1963, 199, and the website: http://www.pensees-depascal.fr/Transition/Transition4-moderne.php.

[2] *Principles of Philosophy*, part I, article 23; CSM 201. Descartes also derives the laws of motion and the conservation of the quantity of motion in the universe from the eternal and immutable nature of God. See *Principles of Philosophy*, Part II, article 36; CMS 240.

regards as indefinitely, rather than infinitely, small. In a similar spirit, Leibniz and Spinoza both distinguish God's infinity from the kind of infinity proper to mathematics. And in contrast, Newton's view of space as an emanative effect of God stems from Newton's desire to reconcile the infinity of space with God's status as the infinite being.

The finitude of the human mind left Pascal wondering how human beings could ever hope to understand the infinite. Indeed, he even argues that any such hope would be presumptuous. But other early modern thinkers attempted exactly that, utilizing, adapting, and modifying a nuanced, technical vocabulary in their attempt to understand infinity. For example, Leibniz claims that mathematical considerations concerning infinity prove extremely useful in philosophy. In a well-known passage, he writes:

> At last a certain new and unexpected light shined from where I least expected it, namely from mathematical considerations on the nature of infinity. For there are two labyrinths of the human mind, one concerning the composition of the continuum, and the other concerning the nature of freedom, and they arise from the same source, infinity. (Leibniz 1989, 95)

Early modern attempts to grapple with infinity tended to recognize and discuss three basic distinctions regarding the infinite – distinctions with roots in the Scholastic philosophical tradition. First, many early modern philosophers recognized – though sometimes only implicitly – a distinction between qualitative and quantitative conceptions of infinity. This distinction was utilized by philosophers like Leibniz and Spinoza, who both distinguish God's qualitative infinity from the quantitative conception of infinity considered in mathematics. Second, early modern philosophers acknowledged a distinction between actual and potential infinity. The Scholastic tradition standardly rejected the existence of actual infinities in the created world but, following Aristotle's approach to Zeno's paradoxes, embraced potential infinities. One finds early modern philosophers like Locke embracing the concept of a potential infinity in his claim that our representation of space is such that one can always imagine more and more space than one has already imagined at any given time but that one cannot imagine the entirety of space. However, other early modern philosophers argued against the Scholastic tradition and claimed that actual infinities are, in fact, possible in the created world – for instance, Leibniz maintains that "instead of admitting that nature rejects it (actual infinity), as it is vulgarly said, I hold that it affects it everywhere, for better marking the perfections of its author".[3] Third, early modern thinkers were also aware of the distinction between the categorematic and syncategorematic concepts of infinity. Although medieval and early modern authors sometimes identified the actual/potential infinity and categorematic/syncategorematic infinity dichotomies with one another, various recent commentators have noted that these two dichotomies are, in fact, distinct. For example, some commentators suggest that a categorematic infinity is one that is larger than any finite number, whereas a syncategorematic infinity is a collection

[3] Leibniz to Foucher in Leibniz, G.W. (1978, 416).

such that, for any member, there exists a larger member. And this distinction is independent from whether the members of a collection are actual or potential. Indeed, one philosopher who keenly recognized this fact was Leibniz, who maintains that there are actual syncategorematic infinities.[4]

As we can see, early modern philosophers employed various conceptions of the infinite in a wide variety of mathematical, scientific, and theological contexts. The aim of this volume is to illustrate the richness of early modern philosophical thought regarding infinity and thus to enhance and facilitate the further study of infinity in this period. In particular, the contributions to this volume aim to demonstrate the different conceptions of infinity recognized by early modern philosophers and the varied uses to which these different conceptions of infinity were put. We do not think that any volume on the topic of infinity in early modern philosophy can claim to be comprehensive, and we make no such claims for this volume. Instead, the volume focuses in large part on three figures for whom infinity played an especially crucial role – namely, Descartes, Leibniz, and Spinoza – and concludes with Kant, who is traditionally considered the final major figure in early modern philosophy. The remainder of this introduction contains the authors' abstracts of their contributions.

In his chapter, Igor Agostini presents some important facets of the scholastic background to Descartes's conception of infinity. In particular, this chapter, considers Francisco Suárez's role in the late medieval debate over the concept of the relationship between God's status as a perfect being and God's status as an infinite being. Although Agostini does not argue that Descartes knew Suárez's position when he originally wrote the *Meditations*, he shows that Suárez's position lies behind Caterus's criticisms of Descartes in the *Objections and Replies* that Descartes published with the *Meditations*. Thus, knowledge of Suárez's position is important for assessing both the reception and success of Descartes's proofs of God's existence in the *Meditations*.

In her chapter, Anat Schechtman notes that Descartes's metaphysics posits a sharp distinction between two types of non-finitude, or unlimitedness: whereas God alone is infinite, numbers, space, and time are indefinite. As Schechtman observes, this distinction has proven difficult to interpret in a way that abides by the textual evidence and conserves the theoretical roles that the distinction plays in Descartes's philosophy—in particular, the important role it plays in the causal proof for God's existence in the *Meditations*. After formulating the interpretive task, Schechtman criticizes extant interpretations of the distinction. She then proposes an alternative at whose core is the idea that whereas the indefinite is a structural, iterative notion, designating the absence of an upper bound, the infinite is an ontic notion, signifying being in general, or what is, without qualification.

In his chapter, Geoffrey Gorham considers how in two rarely discussed passages – from unpublished notes on the *Principles of Philosophy* and a 1647 letter to Chanut – Descartes argues that the question of the infinite (or indefinite) extension of space is importantly different from the infinity of time. In both passages, Descartes

[4] See Richard Arthur's contribution to this volume.

is anxious to block the application of his well-known argument for the indefinite extension of space to time, in order to avoid the theologically problematic implication that the world has no beginning. Descartes concedes that we always imagine an earlier time in which God might have created the world if he had wanted, but insists that this imaginary earlier existence of the world is not connected to its actual duration in the way that the indefinite extension of space is connected to the actual extension of the world. This chapter considers whether Descartes's metaphysics can sustain this asymmetrical attitude towards infinite space vs. time. Gorham first considers Descartes's relation to the 'imaginary' space/time tradition that extended from the late scholastics through Gassendi and More. He then examines carefully Descartes's main argument for the indefinite extension of space and explains why it does not apply to time. Most crucially, since duration is merely conceptually distinct from enduring substance, the end or beginning of the world entails the end or beginning of real (as opposed to imaginary or abstract) time. In contrast, extension does not depend on any enduring substance besides itself.

In his chapter, Yitzhak Melamed notes that Spinoza defines God at the opening of the *Ethics* as a substance consisting of infinitely many attributes. Still, as Melamed observes, the reader of the *Ethics* will find only two of these attributes discussed in any detail in Parts Two through Five of the book. Addressing this intriguing gap between the infinity of attributes asserted in E1d6 and the discussion of merely the two attributes of Extension and Thought in the rest of the book, Jonathan Bennett writes: "Spinoza seems to imply that there are other [attributes] – he says indeed that God or Nature has 'infinite attributes.' Surprising as it may seem, there are reasons to think that by this *Spinoza did not mean anything entailing that there are more than two attributes*." In this chapter, Melamed shows that Bennett's claim is fundamentally wrong and deeply misleading. He does think, however, that addressing Bennett's challenge helps us better understand Spinoza's notion of infinity. He begins by summarizing Bennett's arguments and then turns to examine briefly the textual evidence for and against his reading. He responds to each of Bennett's arguments, and concludes by pointing out some theoretical considerations that, he believes, simply refute his reading.

In her chapter, Sanja Särman notes that Spinoza's widely debated understanding of the infinite has not figured prominently in accounts of his therapy of the passions, and sets out to explore this topic. Her reasoning consists of six steps. First, she introduces Spinoza's cognitive therapy, which claims that we can be healed from our passions by acquiring adequate ideas of them. Second, she shows that Spinoza's cognitive therapy is subject to the Affective Force Condition, i.e., the claim that only a stronger affect can defeat another affect. Third, she suggests that the one way an affect can be reliably weakened or strengthened is by us considering its intentional object as necessary or free, respectively. Fourth, she argues that we can only adequately ascribe freedom to things if we regard them as absolutely infinite, and necessity to finite things if we can have an adequate idea of them as inscribed in an infinite causal nexus. This presupposes having an adequate idea of the infinite; and so fifth, she asks by which capacity, or from which perspective, we can acquire such an idea, on Spinoza's view. Sixth, she finally asks whether the infinity of the causal history of a finite thing qualifies as the kind of infinity which we can adequately conceive, and gives a tentative (affirmative) answer to this question by situating this infinity in Spinoza's taxonomy of infinities.

In her chapter, Noa Shein acknowledges that there are many aspects of the *Ethics* that seem to suggest, or perhaps even require the possibility of deducing finite modes from the infinite substance. Nevertheless, as Shein also observes, many have noted even during Spinoza's own time, that it is far from clear that such a deduction can be successfully performed. Shein argues that the expectation of a top-down deduction (i.e., one that begins with an attribute and terminates with a finite mode) is unwarranted, and that interestingly enough, it is not only unwarranted with regard to Spinoza but with regard to Descartes as well. She shows this by pointing to the crucial role confusion plays for both, noting that our epistemic journey to clear and distinct or adequate knowledge begins with confusion, and that this trajectory is one of emending initially confused ideas. This shows that epistemically a bottom-up trajectory necessarily precedes a top-down one. Her claim, however, goes beyond this point regarding the order of discovery. She argues that the state of confusion presupposes a plurality of finite modes, in Spinoza's case, and substances and modes in Descartes's. Confusion, she claims, amounts to an inadequate perception of an amalgam of ideas of a real plurality of things. Recognizing the metaphysical ground of confusion along with its central epistemic role brings to light the artificiality, and indeed ultimately the impossibility of a top-down deduction which is divorced from a preceding bottom-up analysis.

In his chapter, Mogens Lærke examines how in 1697, the publication of a letter from Leibniz to Bourguet in the *Journal des Sçavants* prompted a vigorous reply from the Cartesien Pierre-Sylvain Regis, leading to a subsequent public exchange between the two philosophers. The controversy ended with a contribution by Regis who seemingly got the final word. The exchange mainly focused on Descartes's *Principles of philosophy*, III, art. 47, a text where Descartes held that the world would eventually take all the possible forms of which it is capable. While Leibniz maintained that this claim had inescapable Spinozistic consequences, Regis defended Descartes against the accusation. Each position relied on its own distinct conception of the world's infinity. This chapter follows this controversy step by step. Laerke shows in particular that Leibniz, although he did not write it, had a forceful reply to Regis's final published objections ready at hand, based on his rehabilitation of the notion of actual infinity.

Ohad Nachtomy's chapter seeks to highlight some of the main threads that Leibniz used in developing his views on infinity in his early years in Paris (1672–1976). In particular, Nachtomy focuses on Leibniz's encounters with Descartes, Galileo, and Spinoza. Through these encounters, some of the most significant features of Leibniz's view of infinity begin to emerge. Leibniz's response to Descartes reveals his positive attitude to infinity. He rejects Descartes's view that, since we are finite, we cannot comprehend the infinite and therefore should refrain from studying it. Likewise, Leibniz rejects Descartes's view that the term 'infinite' should be reserved for God alone, as well as Descartes's distinction between the infinite and the indefinite. Leibniz's encounter with Galileo brings out his rejection of infinite number in response to Galileo's paradox. This, in turn, leads him to face another formidable challenge, viz., to defend the claim that an infinite being is possible, while an infinite number is not. Leibniz's encounter with Spinoza, Nachtomy suggests, highlights the way he approaches this problem by distinguishing between quantitative and non-quantitative senses of infinity. The strategy of employing different senses of infinity in different contexts will remain central in Leibniz's approach to infinity for the rest of his career.

In his chapter, Richard T. Arthur observes that it is well known that Leibniz advocated the actual infinite, but that he did not admit infinite collections or infinite numbers. However, Arthur also notes that Leibniz's assimilation of this account to the scholastic notion of the syncategorematic infinite (more accurately, the infinite syncategorematically understood) has given rise to controversy. A common interpretation is that in mathematics Leibniz's syncategorematic infinite is identical with the Aristotelian potential infinite, so that it applies only to ideal entities, and is therefore distinct from the actual infinite that applies to the actual world. Against this, Arthur argues that Leibniz's actual infinite, understood syncategorematically, applies to any entities that are actually infinite in multitude, whether numbers, actual parts of matter, or monads. It signifies that there are more of them than can be assigned a number, but that there is no infinite number or collection of them (the categorematic infinite), which notion involves a contradiction. Similarly, to say that a magnitude is actually infinitely small in the syncategorematic sense is to say that no matter how small a magnitude one takes, there is a smaller magnitude, but there are no actual infinitesimals. In geometry one may calculate with expressions apparently denoting such entities, on the understanding that they are fictions, standing for variable magnitudes that can be made arbitrarily small, so as to produce demonstrations that there is no error in the resulting expressions.

In his chapter, Paul Guyer argues that Kant's claim in the "Transcendental Aesthetic" of the *Critique of Pure Reason* that space and time are immediately given in intuition as infinite magnitudes is undercut by his general theory of mathematical knowledge. On this general theory, pure intuition does not give objects of any determinate magnitude at all, but only *forms* of possible objects. Specifically, what pure intuition itself yields is the recognition that any determinate space or time is part of a larger one, but it requires an inference of reason to go from that to the claim that space and time are infinite. Guyer further argues that this result is consistent with Kant's claim in the second-edition "Transcendental Deduction" that the unity of space and time are the products of synthesis, but also means that the unity of space and time as objects cannot be used a *premise* in the Transcendental Deduction but can only be regarded as a *conclusion* of the Deduction and the following "System of Principles."

The editors wish to clarify that the each chapter is self-contained. For this reason, there are some slight variations between the chapters with respect to the bibliographical list and the method of reference.

Abbreviations

CSM Descartes, René. (1984–5). *The Philosophical Writings of Descartes*. Vols. 1-2. Ed. and trans. J. Cottingham, R. Stoothoff, and D. Murdoch. Cambridge: Cambridge University Press.

References

Euler, L. (1833). *Letters of euler on different subjects in natural philosophy addressed to a German princess with notes, and a life of Euler*. (Vol. 2, D. Brewster, Trans.). New York: Harper & Brothers.

Leibniz, G. W. (1978). In C. I. Gerhardt (Ed.), *Die philosophischen Schriften von Leibniz* (Vol. 7, pp. 1875–1890). Berlin/Hildesheim: Weidmann/Olms.

Leibniz, G. W. (1989). In R. Ariew & D. Garber (Eds.), *Leibniz: Philosophical essays*. Indianapolis: Hackett.

Leibniz, G. W. (2003). *Double infinity in pascal and monad* (L. Strickland, Trans.). Leibniz Translations. http://www.leibniztranslations.com/pascal.htm. Last Accessed 6 July 2017.

Pascal, B. (1995). *Pensées and other writings* (H. Levi, Trans.). Oxford: Oxford University Press.

Chapter 2
Descartes's *ens summe perfectum et infinitum* and its Scholastic Background

Igor Agostini

Abstract This chapter presents some important facets of the scholastic background to Descartes's conception of infinity. In particular, this chapter considers Francisco Suárez's role in the late medieval debate over the concept of the relationship between God's status as a perfect being and God's status as an infinite being. Although I do not argue that Descartes knew Suárez's position when he originally wrote the *Meditations*, I show that Suárez's position lies behind Caterus's criticisms of Descartes in the *Objections and Replies* that Descartes published with the *Meditations*. Thus, knowledge of Suárez's position is important for assessing both the reception and success of Descartes's proofs of God's existence in the *Meditations*.

2.1 Introduction

The final and decisive step of the first proof of God's existence in Descartes's Third Meditation is very simple and clear: I have the idea of God, that is the idea of a substance that is infinite, eternal, immutable, independent, supremely intelligent, supremely powerful, and which created both myself and everything else (if anything else there be) that exists; now, the idea of an infinite substance (and this idea only) cannot be caused by myself, insofar as I am a finite substance; therefore, an infinite substance must exist as a cause of this idea (CSM II 31, AT VII 45).

However, in spite of the simplicity and clearness of the proof's core, some aspects remain puzzling. Among them is the relationship of the divine name *infinitum* with the other divine name *perfectum*: indeed, though the core of the proof turns on the idea of an infinite substance, in order to prove that this idea is a true one (*idea vera*), Descartes presents it as the idea of a supremely perfect and infinite being (*ens summe perfectum et infinitum*) (CSM II 31, AT VII 46). As a result, in just a few

I. Agostini (✉)
Department of Humanistic Studies, University of Salento, Lecce, Italy
e-mail: igor.agostini@unisalento.it

O. Nachtomy, R. Winegar (eds.), *Infinity in Early Modern Philosophy*, The New Synthese Historical Library 76,
https://doi.org/10.1007/978-3-319-94556-9_2

lines, the text shifts from the idea of an infinite substance to the idea of a supremely perfect and infinite being.

A question arises here: even if we consider *substantia* and *ens* to be synonymous, or at least equivalent, can the two names *infinitum* and *summe perfectum* be considered synonymous too? The Third Meditation suggests that this is the case, because Descartes seems to move easily and unproblematically from one to the other. It should not surprise us, therefore, that a long tradition of interpreters has ascribed to Descartes the thesis of an identity between infinity and perfection.[1]

Nevertheless, if we consider things from a broader perspective, not restricted to the text of the Third Meditation, some doubts emerge. Indeed, Descartes's theory of the creation of the eternal truths entails the idea that every order, law, truth, and good (*nullum ordinem, nullam legem, nullamve rationem veri et boni*), and so every mathematical and metaphysical truth (*tam Mathematicae quam Metaphysicae veritates*) depends on God's immensity (CSM II 293–294, AT VII 435–436).[2] Accordingly, if we assume that perfection belongs to the domain of the metaphysical truths dependent on God, and that God's immensity coincides with his infinity,[3] it follows that perfection is subordinated to infinity. This is confirmed by the fact that although Descartes repeats quite obsessively that infinity is incomprehensible (including the canonic text on the clearness and distinction of the idea of God: CSM II 31, AT VII 46), he scarcely makes the same claim for perfection.[4] Moreover, as is well known, the proof of God's existence in the Fifth Meditation, which is based on an analogy with mathematical demonstrations (that is, the domain of what is comprehensible), is not textually grounded upon God's infinity, but on the notion of God as *ens summe perfectum* (CSM II 45, AT VII 65).

For this reason, some authors have recently argued that the two names of infinity and perfection, far from being synonyms, express two different determinations of Descartes's definition of God.

Such an interpretation, which was previously advanced, though not systematically elaborated, by authors like Octave Hamelin and Jacques Maritain, has been more recently re-proposed and developed by Jean-Luc Marion, who traces the opposition of infinity and perfection back to an alleged contrast between two

---

[1] Liard, *Descartes* (Paris: Alcan, 1903), 190; Alexandre Koyré, *Essai sur l'idée de Dieu et les preuves de son existence chez Descartes* (Paris: Leroux, 1922), 140; Étienne Gilson, *L'esprit de la philosophie médiévale* (Paris: Vrin, 1932; 3rd ed. 1969), 58; Henri Birault, *Essai sur les attributs divins chez Descartes. Mémoire pour l'obtention du Diplôme d'études supérieures. Sous la direction de Monsieur Henri Gouhier* (Paris: E.N.S., 1939), 38; Virgilio Lazzeroni, *La formazione del pensiero cartesiano e la Scolastica* (Padova: Cedam, 1940), 181; Henri Gouhier, *La pensée métaphysique de Descartes* (Paris: Vrin, 1962), 191; Cornelio Fabro, *Introduzione all'ateismo moderno* (Roma: Editrice Studium, 1964), 38; Jean-Marie Beyssade, *La philosophie première de Descartes. Le temps et la cohérence de la métaphysique* (Paris: Flammarion, 1979), 313.

[2] See also CSM II 291–292, AT VII 432.

[3] At least, for the reason that it is from the attribute of God's infinity that Descartes argues for the creation of the eternal truths. See also, AT I 150, CSMK III 24–5 and AT I 152, CSMK III 25.

[4] One exception is *Meditatio III*: "omnes illas perfectiones, quas ego non comprehendere, sed quocunque modo attingere cogitatione possum" (AT VII 51–52, CSM II 35).

different theological traditions: Ockhamism (infinity) and Scotism (perfection). The idea of an opposition between the two attributes of infinity and perfection had a certain success with interpreters, especially in France,[5] but it was also strongly criticized by other scholars such as Jean-Marie Beyssade, Philip Clayton, and others.[6] In particular, as concerns the historical background, Aza Goudriaan, in his book on Descartes and Suárez's knowledge of God, contested Marion's alleged opposition between infinity and perfection in the theological tradition before Descartes by arguing, on the contrary, that Descartes identifies infinity with perfection in the wake of Suárez.

Nevertheless, in order to grasp the exact sense of Suárez's position, the identification that he actually (as I will show) establishes between infinity and perfection should not be assimilated[7] to the traditional notion of infinity as perfection, which does not denote a formal identification between infinity and perfection, but only expresses the point that God's infinity has to be understood as an entitative and perfect infinity, and consequently, sharply distinguished from quantitative forms of infinity. This is the source of the traditional distinctions between infinity *secundum essentiam*, or *secundum perfectionem*, and *secundum magnitudinem*, or *secundum quantitatem*, and so on, elaborated by Christian theologians since the times of the Church Fathers, and then endorsed by the Scholastics,[8] in order to attribute infinity

---

[5] See Jean-Luc Marion, *Sur le prisme métaphysique de Descartes* (Paris: PUF, 1986), 276 ss.; Pierre Magnard, "La fonction 'Dieu' dans le système," *Revue des sciences Philosophiques et Théologiques*, LXXII (1988), n. 2, 195–207; Vincent Carraud, "Descartes appartient-il à l'histoire de la métaphysique," in *Lire Descartes aujourd'hui*, ed. Olivier Depré et Danielle Lories (Louvain-la-Neuve/Paris: Institut Supérieur de Philosophie/Peeters), 1997, 153–171; and, for certain respects, Jean-Louis Chedin, "Infini et subjectivité dans la pensée classique," in *Revue de Métaphysique et de Morale*, XCIV (1989), n. 2, 229–250. Among the old studies, see Émile Boutroux, *De veritatibus aeternis apud Cartesium.* Haec apud Facultatem litterarum Parisiensem disputabat Em. Boutroux (G. Parisiis: Baillière, 1874), French translation by Georges Canguilhelm, (Paris: Alcan, 1927), 59 (though the opposition is here between perfection and infinity of divine power); Octave Hamelin, *Le système de Descartes* (Paris: Alcan, 1921 [1911]), 227 ss. (speaking of a mutual subordination between the two attributes); Jacques Maritain, *Le songe de Descartes. Suivi de quelques Essais* (Paris: Buchet/Castel, [s. d.]) 240–241 and 328, n. 48. Cfr. 227–228.

[6] See Philip Clayton, "Descartes and Infinite Perfection," *Proceedings of the American Catholic Philosophical Association*, LXVI (1992), 137–147; "The Theistic Argument from Infinity in Early Modern Philosophy," *International Philosophical Quarterly*, XXXVI (1996), n. 1, 5–17; *Das Gottesproblem. Band 1: Gott und Unendlichkeit in der neuzeitlichen Philosophie* (Paderborn/München/Wien/Zürich: Schöningh, 1996), 43–101; Jean-Marie Beyssade, D*escartes au fil de l'ordre,* (Paris: PUF, 2001); Aza Goudriaan, *Philosophische Gotteserkenntnis bei Suárez und Descartes: im Zusammenhang mit der nederländischen reformierten Theologie und Philosophie des 17. Jahrhunderts* (Leiden-Boston-Köln: Brill, 1999), 158, 165, 313. More recently, see Laurence Devillairs, D*escartes et la connaissance de Dieu* (Paris: Vrin, 2004), 50 ss.; Igor Agostini, L*'idea di Dio in Descartes. Dalle M*editationes *alle* Responsiones (Firenze-Milano: Le Monnier-Mondadori Education, 2010), 30, 40–41; Dan Arbib, D*escartes, la métaphysique et l'infini* (Paris: Vrin, 2017), 156–159.

[7] This is what Goudriaan, *Philosophische Gotteserkenntnis*, 237, seems to do when he explains Descartes's identification between infinite and perfect on the basis of the exigence of distinguishing God's infinity from the infinity of the world. See, also, *infra*, p.

[8] See, for example, Thomas de Aquino, *Summa theologiæ*, I, 7, 3, c.: "Respondeo dicendum quod aliud est esse infinitum secundum suam essentiam, et secundum magnitudinem".

to God and to go beyond the Greek opposition between infinity and perfection. From the time of Augustine onward, all Christian theologians have agreed that God's infinity has to be understood as infinity of perfection.[9]

However, once we have established this point, a different question arises: in what exactly does the nature of divine infinity consist? Or, in other words, what constitutes the nature of infinity *secundum essentiam* or *secundum perfectionem*? Such a question was in fact not important to the Medieval theologians, who were much more concerned with demonstrating that God is infinite (in the sense of an entitative infinity) than in determining the nature of this infinity. The case of Thomas Aquinas is telling here: in the seventh question of the First Part of *Summa theologiae*, dedicated to the infinity of God (*De infinitate Dei*), he only addresses the question *If God is infinite* (Article 1: *Utrum Deus sit infinitus*). Of course, there were some exceptions, like John Duns Scotus,[10] but what is certain is that in Medieval theology the question of the nature of God's infinity never developed into a debate. Nevertheless, at the end of the sixteenth century, after the publication of Francisco Suárez's (1548–1517) *Metaphysical Disputations* (1597), things changed: a new debate, which originated with Suárez, concerning the *ratio formalis* (that is, the nature of God's infinity) quickly became one of the most intense controversies in Early Modern Scholastic theology. It is precisely within this controversy (which is still not very well known) that Suárez's conception of God's infinity has to be understood.[11] The meaning, both historical and conceptual, of Suárez's identification of infinity and perfection becomes much clearer if understood in light of this controversy. So, in addressing the problem of the Scholastic background of Descartes's notion of *ens summe perfectum et infinitum*, it is impossible to ignore this debate, even if only to conclude that, both historically and conceptually, Descartes's notion cannot be explained in the light of it.

In this paper, I will proceed in three steps. First, I will sketch the main points of the debate concerning the *ratio formalis* of God's infinity. Second, I will establish a conceptual comparison between the notion of *ens summe perfectum et infinitum* in Descartes's Third Meditation and Suárez's concept of divine infinity. Third, I will address, from a strictly historical point of view, the problem of the eventual

[9] For the history of the infinity of God, see Leo Sweeney, *Divine Infinity in Greek and Medieval Thought* (New York: Peter Lang Publishing, 1992). For the Medieval Age, in particular, see Antoine Coté, *L'infinité divine dans la théologie médiévale (1220–1255)* (Paris: Vrin, 2002); Anne A. Davenport, *Measure of a Different Greatness. The Intensive Infinite*, 1250–1650, (Leiden-Boston-Köln: Brill, 1999).

[10] See, in particular, the discussion of the intensive infinity in Question Five of his *Quodlibeta, utrum relatio originis sit formaliter infinita* (Johannes Duns Scotus, *Quaestiones quodlibetales*, q. 5, a. 1, in *Opera omnia quae hucusque reperiri potuerunt, collecta, recognita, notis, scholis et commentariis illustrata a P. P. Hibernis Collegii romani S. Isidori professoribus*, ed. L. Wadding, 12 voll. (Lugduni: sumptibus Laurentii Durand, 1639), rist. anast. (Hildesheim: Olms, 1968–1969) ed. Lugduni, cit., vol. XII, 113–141, especially 119).

[11] This is the main point of my book, *L'infinità di Dio. Il dibattuto da Suárez a Caterus (1597–1641)* (Roma: Editori Riuniti University Press, 2008).

relationship between Descartes's notion of *ens summe perfectum et infinitum* and the debate on the *ratio formalis* of God's infinity.

2.2 The Scholastic Debate on the Ratio Formalis of God's Infinity

What I call the debate concerning the *ratio formalis* of God's infinity is a controversy that arose at the end of the sixteenth century, and mainly developed in the commentary on the *Summa theologiae* on the nature of God's infinity. This controversy, which originated with Suárez's discussion of God's infinity in his *Metaphysical Disputations*, develops through the intervention of Gabriel Vázquez's (1549–1604) *Commentaries* of 1598, and expands with Suárez's reply in the first part of his enormously influential *Treatise on Divine Substance*, published in 1606.

The debate reached its acme at the mid-seventeenth century and involved all the greatest Scholastic authors of the time as well as most of the minor authors. Though it is impossible to reconstruct this debate here, I will sketch the positions of three principal groups engaged in the controversy: (1) Suárez, who identified infinity with perfection; (2) The Thomists, who identified infinity with divine simplicity and, in particular, with the fact that the divine being is not received in the essence[12]; and (3) Vázquez, who identified infinity with the claim that the divine being is *a se* (what can be called *aseitas*, or aseity).

In this debate, Suárez's solution prevailed, though he developed his ideas only gradually. Suárez addressed the question for the first time in the *Metaphysical Disputations*, where he states that infinity consists in the fact that the perfection of the first being, unlike finite beings: (1) is not limited to one genus only but includes *eminenter* all genera; and (2) is not limited to a certain degree in each genus.[13]

Suárez was not simply claiming that God's infinity has to be distinguished from quantitative infinity (according to the traditional distinction between infinity

---

[12] In fact, taken in itself, this identification was far from being made explicit in Aquinas, who simply deduced infinity from the fact that the divine being is not received in the essence, without pretending to identify the two (see, in particular, *Summa theologiae*, I, 7, 1, c.). In this sense, the solution of the Thomists in the seventeenth century was not exactly in compliance with Aquinas; and it is not surprising that their adversaries contested their pretended fidelity to Aquinas or pretended (like Suárez) that their own solution was more faithful to Aquinas.

[13] Francisco Suárez, *Metaphysicae disputationes*, disp. 30, sect. 2, n. 21, in *Opera omnia*, 28 voll. (Parisiis: apud Ludovicum Vivès, 1856–1878), vol. XXVI, 70b–71a: "Haec infinitas in nullo alio consistit, nisi in hoc quod perfectio primi entis, nec est. ita praecisa ac definita ad unum genus perfectionum, quae nos in creaturis distingui videmus, ut illud solum includat, et non caetera omnia, eo eminentissimo modo qui ad summam perfectionem pertinere potest; neque etiam in singulis perfectionum generibus est. ita limitata ad certum aliquem et definitum gradum qui in participato ente intelligi possit, quin habeat perfectionem illam nobiliori et excellentiori modo quam possit a creatura participari, etiamsi magis et magis infinitum participetur. Sed hoc totum includitur in perfectione primi entis, ratione cujus dicitur omnem possibilem perfectionem continere, ut ex dictis in praecedenti assertione constat".

secundum magnitudinem and *secundum perfectionem*). He was affirming that perfection constitutes the nature of God's infinity itself. This means that the two attributes of infinity and perfection, which the former theological tradition (Aquinas included)[14] had considered as two different attributes (at least by means of the so-called *distinctio rationis*), is now seen as identical. This explains why the first of the two proofs of God's infinity given by Suárez is based on the attribute of perfection. Indeed, given the identity between supreme perfection and infinity, the proof of God's infinity coincides with the proof of his perfection:

It is from that same principle that God's infinity and his supreme perfection are inferred, because, on the one side, the concept of perfection, such as explained by us, is convertible with the direct definition of the infinity, and, on the other side, infinity as such belongs to perfection and, therefore, to the supreme perfection of being.[15]

The second proof is premised on the assumption that God has no cause, and it runs as follows: every finite being can be limited only by a cause; now, God has no cause; therefore, there is no principle of his limitation, neither from another agent nor from elsewhere.[16] In brief, in so far as he is not caused, God has no principle of limitation:

> Insofar as he has no cause of his being, it is not possible that there is a limitation in him.[17]

It should be observed, however, that the first argument is also ultimately grounded on God's aseity (that is the property by which divine being is uncaused), because God's perfection, from which Suárez deduced infinity, is in itself deduced from aseity. This means that Suárez's demonstration of God's infinity rests entirely on the attribute of aseity that, accordingly, completely replaces the role played in Aquinas by the subsistence of God's being (that is the property by which divine being is identical with its own essence and it is not received in it).

However, in establishing in the *Metaphysical Disputations* that infinity could be deduced from aseity, Suárez did not explicitly point out that aseity, unlike perfection, did not constitute the *nature* of God's infinity; he did not detail the relationship between aseity and perfection, and did not formulate the question of the *ratio*

---

[14] In fact, in the *Prima pars* of his *Summa theologiae*, Aquinas deals separately with God's perfection (Question 4) and his infinity (Question 7).

[15] Suárez, *Metaphysicae disputationes*, 71a: "Ergo ex illo principio aeque infertur infinitas, sicut et summa perfectio, quia et modus perfectionis a nobis expositus convertitur cum infinitate recte declarata, et infinitas, quatenus sub illa negatione perfectionem indicat, pertinet ad perfectionem simpliciter atque adeo ad summam entis perfectionem".

[16] Suárez, *Metaphysicae disputationes*, n. 22, 71a: "Secundo potest probari eadem illatio quia esse per essentiam non habet unde limitetur; esse enim participatum limitari potest, aut ex voluntate dantis tantam perfectionem, et non majorem, aut ex capacitate recipientis, sive illa capacitas intelligatur per modum passivae potentiae, sive tantum per modum objectivae, seu non repugnantiae; in primo autem ente, quod ex se est. suum esse, nullum principium aut ratio limitationis intelligi potest".

[17] Suárez, *Metaphysicae disputationes*, n. 22, 71a: "Quia sicut nullam habet causam sui esse, ita non potest in illo habere limitationem".

formalis of God's infinity. To my knowledge, the first theologian who explicitly addressed this question was Vázquez in his *Commentaries* on the *Summa theologiæ*, which were published in 1598. Here, Vázquez both criticizes Suárez and the followers of Aquinas arguing that infinity should be identified with neither perfection nor with the subsistence of God's being. Indeed, according to Vázquez, what constitutes the *ratio formalis* of God's infinity is the very fact that God has no cause of his own being.[18]

Vázquez's intervention offered Suárez the possibility to return to the question in his *Treatise on Divine Substance*, published in 1606, where, endorsing Vázquez's approach, he explicitly poses the question of the *ratio formalis* of God's infinity. This gives him the opportunity to clarify his own position on the relationship between infinity and aseity. Though he maintains the point he made in his *Metaphysical Disputations*, that infinity can be deduced from aseity, he now points out that they are not formally identical. Aseity, indeed, excludes formally and directly dependence, which is not the case with infinity. In particular, he claims that it is not formally contradictory to think of an angel that is not caused, and nevertheless finite, and on the other hand, an infinite being caused by God.[19]

On this basis, Suárez offers his own solution by establishing that it is in perfection that the *ratio formalis* of divine infinity has to be located:

> God is said infinite absolutely, properly and *quasi a priori*, in so far as he has so much perfection that it is impossible that a being of different nature has more or equal perfection. Accordingly, I assume that the meaning of the infinity consists in the very negation by which many saints explain what God is, i.e. that he is the being a greater than which cannot be conceived.[20]

Suárez's solution to the problem of the *ratio formalis* of God's infinity will dominate the development of the debate and will penetrate the ideas of a great number of authors, unsatisfied by the alternative accounts proposed by Vázquez, on the one

---

[18] Gabriel Vázquez, *Commentaria ac disputationes in primam partem S. Thomae* (1598), 2 voll. (Ingolstadii: Martini Nutti et Ioannis Hertsroy, 1609), pa. 1, q. 7, a. 2, disp. 25, c. 5, vol. I, 144a: "Propria igitur ratio infinitatis Dei secundum essentiam in eo posita est, ut a nulla causa habeat esse".

[19] Francisco Suárez, *Tractatus de divina substantia*, lib. 2, c. 1, n. 2, in *Opera omnia*, ed. Vivès, vol. I, 46ab: "Denique esse a se immediate solum negat dependentiam a causa extrinseca, esse autem infinitum formaliter hoc non negat: nam si mente concipietur angelus finitae naturae et ex se existens sine creatione, formaliter non apprehenduntur duo contradictoria, licet virtualiter ibi includi per discursum ostendi possit. Neque e converso est aperta et formalis contradictio, si apprehendatur ens infinitum a Deo factum, unde quaeri solet an id fieri possit, ut statim dicam: non est ergo negatio, quam dicit infinitas, formaliter eadem cum negatione essendi ab alio".

[20] F. Suárez, *Tractatus de divina substantia*, lib. 2, c. 1, n. 5, in *Opera omnia* cit., vol. I, 47a: "Deus ergo dicitur infinitus simpliciter proprie et quasi a priori, quia tantae perfectionis est, ut non possit in ea habere superiorem nec aequalem, qui sit distinctae naturae. Unde quod multi sancti explicant quid sit Deus, per hoc *quod est tale ens, quo majus excogitari non potest*, hanc ipsam negationem per infinitatem significari intelligo".

hand, and the Thomists, on the other. Indeed, Suárez's identification between infinity and perfection will become the prevalent account of God's infinity in early modern scholastic theology.[21]

This is, very briefly, the historical context that precedes Descartes's elaboration of the notion of *ens summe perfectum et infinitum*. However, in order to better understand the sense of the Suárezian identification between infinity and perfection, something else has to be added.

As we have seen, Suárez affirms that God is said to be infinite absolutely, properly and *quasi a priori*. But what does *quasi a priori* mean? The text is not explicit, which can be explained, as often happens in the *Treatise of the Divine Substance*, by the fact that Suárez had already addressed the point in the *Metaphysical Disputations*. In fact, there is a passage in the Thirtieth Disputation that clarifies the puzzling expression *quasi a priori*. Here, Suárez answers a possible objection against the deduction of infinity from aseity, claiming that the first being could be limited by himself (*ex se*), without any cause, to a certain genus or degree of perfection:

> You will claim that as the first being is by himself, he could be limited by himself, and by no other cause, to a certain genus and degree of perfection.[22]

As we will see, this is precisely the objection that Caterus will advance against Descartes in the *First Replies*. However, before coming to the *Meditations*, let us focus a bit more on Suárez. His solution is that the possibility presented in the objection contradicts the notion of being *necessarium ab intrinseco*. This can be proven – Suárez argues – *a posteriori* from the disadvantages that would follow, first of all that it will not be contradictory that many beings *a se* exist. The same conclusion can be demonstrated also a priori, but, in this case, the only possible proof is *per non repugnantiam*, that is by negation of each cause of limitation of God's being. But what does it mean that this is *only* a proof *per non repugnantiam*? It means that this is not a demonstration by a true cause and reason of the infinite, but only by the negation of a cause of limitation. Suárez is thus explicit here: it is impossible to give a cause or a *ratio* of God's infinity, because in God there is no cause. The person who would pretend to look for a cause or reason of God's infinity will be voluntarily *pertinax*:

> And, unless someone wants to be voluntarily pertinacious (*voluntarie pertinax*), this is a really sufficient demonstration [...] because it is impossible to find here another *a priori* reason by a positive cause, insofar as such a being has no cause.[23]

---

[21] Here, it is impossible for me to develop this point in a more detailed explanation: see, on the subject, Agostini, *L'infinità di Dio*, 163 ss.

[22] Suárez, *Metaphysicae disputationes*, disp. 30, sect. 2, n. 22, in *Opera omnia*, ed. Vivès, vol. XXVI, 71a: "Dices, sicut primum ens ex se est., ita ex se, sine alia causa esse posse limitatum ad certum genus vel gradum perfectionis".

[23] Suárez, *Metaphysicae disputationes*, disp. 30, sect. 2, n. 21, 71a: "Et revera est haec sufficiens demonstratio, nisi quis velit voluntarie pertinax esse, tum quia hic non potest intercedere alia ratio a priori per causam positivam, cum hoc ens nullam habeat causam".

The meaning of the identification between infinity and perfection is here clarified in a precise sense: if perfection constitutes the *ratio* of God's infinity, this *ratio* cannot be interpreted as a positive cause of God's infinity.

After all, according to Suárez, infinite remains a negative attribute, expressed both linguistically and conceptually by a negation:

> We understand that the *negation* of this limitation is signified by the infinity of the essence attributed to God [...] It is this *negation* that I understand to be signified by this infinity.[24]

As we can see, in spite of his identification between infinity and perfection, Suárez did not abandon at all, at least on this point, the strong restrictions of the traditional negative theology.

2.3 Descartes's Ens Summe Perfectum et Infinitum

An analysis of Descartes's texts shows that his notion of *ens summe perfectum et infinitum* has to be understood precisely in the sense of a formal identity between infinity and perfection. However, in order to make this point, it is important to separate these texts from the others in which Descartes simply distinguishes God's infinity from the indefiniteness of the world as, for example, in this famous passage from his letter to More of April 15th, 1649:

> The reason why I say that the world is indeterminate, or indefinite, is that I can discover no limits in it; but I would not dare to call it infinite, because I perceive that God is greater than the world not in extension (for I have often said I do not think he is strictly speaking extended) but in perfection.[25]

Here, in claiming that God is infinite *ratione perfectionis*, Descartes simply distinguishes God's infinity from the infinity of the world, which is infinite *ratione extensionis*. He does not seem to be concerned with the nature of God's infinity or with an identification between infinity and perfection.[26] The main textual evidence supporting the idea that Descartes establishes a formal identification between infinity and perfection comes from the Third Meditation and, in particular, from the two passages immediately following the final step of his first proof of God's existence, that point out two theses: the idea of God is a *vera idea* and the idea of God is an *idea vera*.

---

[24] Suárez, *Tractatus de divina substantia*, lib. 1, c. 9, 47a: "*Negationem* illius limitationis intelligemus significari per infinitatem essentiae Deo attributam [...] Hanc *negationem* per infinitatem significari intelligo".

[25] *Descartes to More*, 15 April 1649: "Dico idcirco mundum esse indeterminatum vel indefinitum, quia nullos in eo terminos agnosco; sed non ausim vocare infinitum, quia percipio Deum esse mundo maiorem, non ratione extensionis, quam, ut saepe dixi, nullam propriam in Deo intelligo, sed ratione perfectionis" (CSMK 374, AT V 344).

[26] As Goudriaan, *Philosophische Gotteserkenntnis*, 237–238, argues.

These two theses are formulated by Descartes as two answers to two possible objections addressed against his proof. The first objection is that the idea of an infinite substance is not a true idea, because it is formed by negating the finite (CSM II 31, AT VII 45). Here is Descartes's reply to this objection:

> On the contrary, I clearly understand that there is more reality in an infinite substance than in a finite one, and hence that my perception of the infinite, that is God, is in some way (*quodammodo*) prior to my perception of the finite, that is myself. For how could I understand that I doubted or desired – that is, lacked something – and I was not wholly perfect, unless there were in me some idea of a more perfect being which enabled me to recognize my own defects by comparison? (CSM II 31, AT VII 45-46)

In this text it is clear that Descartes considers the concept of infinity as synonymous with the concept of a *maximum* amount of reality; otherwise, this argument would contain a *non sequitur*. Accordingly, given the identity between reality and perfection that Descartes establishes by introducing the causal principle (CSM II 28: "It follows from this both that something cannot arise from nothing, and also that what is more perfect – that is, contains in itself more reality – cannot arise from what is less perfect"), it is clear that he also considers infinity and perfection as synonymous. This is the reason that explains why, in answering the second objection, Descartes can use, for the first time, *summe perfectum et infinitum* (CSM II 31, AT VII 46): the argument for the *vera idea* already presupposes the identity between infinite and perfect and anticipated the notion of *ens summe perfectum et infinitum*.

The second objection is that the idea of God, though a *vera idea*, is materially false and not an *idea vera*, that is coming from nothing, like the idea of the cold (CSM II 31, AT VII 46). In other words, as Descartes later explains in the *First Replies*, the objection is that the idea we form of God does not contain something real, but only a negation, as, for example, the idea of cold contains nothing more than the negation of heat (CSM II 82, AT VII 114). The answer is that the idea of God, far from being devoid of reality, is true in its highest degree, insofar as it contains the maximal amount of reality and perfection. Indeed:

> Whatever I clearly and distinctly perceive as being real and true, and implying any perfection, is wholly contained in it (CSM II 32, AT VII 46).

Once again, the argument would contain a *non sequitur* if an identification between infinity and perfection was not supposed: the answer to the objection that the idea of the infinite contains a negation is grounded on the fact that the content of the idea of the infinite is the quantity (supreme quantity) of reality and perfection.

It should be now clear that any pretended opposition between infinity and perfection contradicts the conceptual foundations of the two crucial theses of the idea of God as *vera idea* and *idea vera*, foundations that presuppose a formal identity between infinity and perfection. In a word, the identification between infinity and perfection is the keystone of the foundation of the two theses of the *vera idea* and *idea vera*.

Moreover, towards the end of the first proof, Descartes even more explicitly asserts the identity between infinity and perfection. In answering the objection that perhaps all the perfections which I attribute to God are somehow potentially in myself, though not actualized, Descartes states that one judges God to be actually infinite, so that nothing can be added to his perfection, "God, on the other hand, I take to be actually infinite (*infinitum*), so that nothing can be added to his perfection (*perfectioni*)" (CSM II 32, AT VII 47). Once again, God's infinity consists precisely in his perfection or, in other words, in that nothing can be added to God's perfection.

However, a very deep difference separates Descartes from Suárez here. According to the latter, as we have seen, the identification between infinity and perfection does not imply a transformation of infinity into a positive attribute. But for Descartes, infinity, far from being a negative notion, is a positive one, as is perfection. The point was made explicitly in claiming that the idea of God is a *vera idea*, "My perception of the infinite [...] is in some way prior to my perception of the finite" (CSM II 31, AT VII 46). What we have here is not simply an identification between infinity and perfection, but a transformation of infinity (and perfection) into a positive attribute. Descartes will strongly insist on this point, though not always coherently, along with Gassendi and the Hyperaspistes:

> It is false that the infinite is understood through the negation of a boundary or limit; on the contrary, all limitation implies a negation of the infinite (CSM II 252, AT VII 365).

> Similarly, I kept the name 'infinite', when 'the greatest being' (*ens amplissimum*) would be more correct if we wanted all names to conform to the nature of the things (CSM III 192, AT III 427).

As in Suárez, perfection constitutes the structure itself (the *ratio*, in Suárezian terms) of God's infinity, but unlike Suárez, according to Descartes, it is a positive structure that determines the idea of God. And it is precisely this structure that we have observed underlying the two theses of the *vera idea* and *idea vera*.

2.4 Is There a Historical Relationship Between Descartes and the Debate on the Ratio Formalis of God's Infinity?

So far so good. However, a methodological question arises at this point: is it possible to establish a historical connection, so that Descartes's notion of *ens summe perfectum et infinitum* depends on Suárez's identification between infinity and perfection? I do not think so. On the basis of the available texts, we cannot establish such a connection. I do not seek to establish that Descartes was aware of the debate concerning the *ratio formalis* of God's infinity, nor to explain his metaphysical choices in light of it, and in particular, in light of Suárez's solution. My aim is more limited: I want to reconstruct the intellectual context and to argue that the identity between infinity and perfection was the most widespread solution to the problem of the *ratio formalis* of God's infinity in Early Modern Scholastic theology. In this sense, by stating an identity between infinity and perfection, Descartes's notion of *ens summe perfectum et infinitum* finds its immediate historical antecedent in the Suárezian doctrine of the infinity of God dominating the Scholastic theology of

Descartes's time. But let us not forget that a historical antecedent is not necessarily a source.

At the same time, it is clear that, at a certain moment, Descartes was asked to situate himself within the debate concerning the *ratio formalis* of the infinity of God, since it reverberated in an extremely interesting passage of the *First Objections* by Johannes Caterus. This happens in discussing Descartes's second proof of God's existence of the Third Meditation, grounded on the notion of being *a se*, that is of a being who "has the power of existing through its own might" and "then undoubtedly it also has the power of actually possessing all the perfections of which it has an idea" (CSM II 34, AT VII 49–50).

According to Caterus, this proof, at most, can prove the existence of a being *a se*, but not its infinity. Indeed, the expression 'from itself' (*a se*) has two senses. In the first, positive sense, it means 'from itself as from a cause'; in the second, negative sense, 'from itself' simply means 'not from another'; and this is the sense – Caterus claims – in which everyone takes the expression *a se*. Now, Caterus is not completely certain about how to interpret Descartes's words, but what is clear to him is that neither in the one sense, nor in the second one, is it possible to prove God's infinity from aseity.

On the one hand, Caterus concedes that the infinity of God is entailed by the first of the two meanings of aseity: "What derives existence from itself in this sense bestows its own existence on itself; so if by an act of premeditated choice it were to give itself what it desired, it would undoubtedly give itself all things, and so would be God" (CSM II 68, AT VII 95). This is a decisive point in order to explain Descartes's reply, as we will see. However, Caterus seems to think that it is not in this sense that Descartes interprets the concept of aseity. Indeed, it is only in the negative sense that everyone has always understood the notion of aseity: "This, as far as I remember, is the way in which everyone takes the phrase in this sense, the notion of aseity" (CSM 68, AT VII 95). Indeed, taken in a positive sense, the notion of aseity is openly contradictory, because it is impossible that a being is the cause of itself: "I shall not listen if you say 'If it derives its existence from itself it could easily have given itself all things'. For it does not derive existence from itself as a cause, nor did it exist prior to itself so that it could choose in advance what it should subsequently be" (CSM II 68–69, AT VII 95).

And, nevertheless, if, on the other hand, we assume the negative sense of aseity, how are we to prove God's infinity? Caterus advances a precise criticism against Suárez. Indeed, he reports an argument that he declares to have borrowed from Suárez:

> Admittedly, I am sure I have heard somewhere that Suárez argued as follows: "Every limitation proceeds from some cause; therefore if something is limited and finite this is because its cause was either unable or unwilling to endow it with more greatness or perfection; and hence if something derives its existence from itself, and not from some cause, it is indeed unlimited and infinite" (CSM II 69, AT VII 95).

And he declares not to agree completely with this argument:

> I do not entirely accept this, however. For what happens if the limitation arises from the thing's internal constitutive principles, that is, from its essence of form? Remember that you have not yet proved this essence to be infinite, even though the thing derives its existence from itself, in the sense of 'not from another'. That which is hot, for example, if you suppose there to be such a thing, will be hot as opposed to cold in virtue of its internal constitutive principles, and this will be true even if you imagine that its being what it is does not depend on anything else (CSM II 69, AT VII 95).

In one word, according to Caterus, a thing could be, at the same time, finite and uncaused, because its limitation could come not from the efficient cause, but from the formal cause. As we have seen, this was also the objection that Suárez, in his *Metaphysical Disputations*, advanced against his own deduction of infinity from aseity: "As the first being is by himself, he could be limited by himself, and by no other cause".[27]

It would be interesting to establish here whether Caterus ignored that Suárez himself had already discussed his objection or – what I think to be more probable – he was simply not persuaded by Suárez's answer. But, in the end, this is not a decisive point. The decisive point is that Caterus's reference to Suárez, far from being a generic claim, is precise and concerns a precise text. This text, which Gilson ignored and has not been clearly identified, not even by Goudriaan, comes from the second of the two demonstrations of the infinity of God contained in Suárez's *Metaphysical Disputations*, which I mentioned previously: "In so far as he has no cause of His being, it is not possible that there is a limitation in Him.[28]

In the heart of the *First Replies*, therefore, one of the most relevant and controversial texts (and doctrines) of the debate on the *ratio formalis* of God's infinity was echoed. And this should not be surprising. Caterus's master, the Belgian theologian Johannes Wiggers (1571–1639),[29] engaged in his commentary to the *Summa theologiae* in the great debate on the infinity of God.[30]

What is even more important, Caterus asks Descartes to measure himself on this point:

> I am sure that M. Descartes has plenty of arguments to support a thesis that others have not perhaps defended with sufficient clarity (CSM II 69, AT VII 95).

Interpreters have always read the objections addressed by Caterus as an expression of an obsolete culture. Now, even if this could be true, at least partially, concerning the passages of the *First Objections* in which Caterus relates Descartes to Aquinas,[31] this is not the case with his quotation of Suárez. The *Primae responsiones* are here updated to the most recent debate of the age. Rather than asking Descartes to deal with the old (though still discussed) Thomistic thesis, Caterus asks

---

[27] See *supra*, section 2.

[28] See *supra*, section 2.

[29] On Caterus as disciple of Wiggers, see the decisive contribution by Jean-Robert Armogathe, "Caterus' Objections to God," in *Descartes and his Contemporaries: meditations, objections, and replies*, ed. R. Ariew et M. Grene (Chicago-London: University of Chicago Press, 1995), 34–43.

[30] Agostini, *L'infinità di Dio*, in particular, 174–176, 187–191, 309–310.

[31] See AT VII 94, 96, 98–99; CSM II 68, 70, 71.

Descartes to offer his own solution to one of the most pressing and controversial theological debates of the time. In this sense, if we are not allowed, as I stated previously, to connect Descartes's doctrine on the infinity of God of the Third Meditation to the debate on the *ratio formalis* of God's infinity, it is certain that, the *First Objections* historically relates Descartes to this debate.

For this very reason, the *First Objections* becomes *ipso facto* a *pièce* in the history of the long polemics on the *ratio formalis* of God's infinity. And, insofar as Descartes, as we will see, agrees to confront Caterus's objection to Suárez, the *First Objections* becomes a *pièce* of this same history, too.

Descartes's answer to Caterus is very well known: "I did not say that it was impossible for something to be the efficient cause of itself" (CSM II 78, AT VII 108). Far from being impossible, a being which is positively *sui causa* is also necessary in order to prove the existence of God. Indeed, Descartes claims:

> If I thought that nothing could possibly have the same relation to itself as an efficient cause to its effect, I should certainly not conclude that there was a first cause. On the contrary, I should go on to ask for the cause of the so-called 'first' cause, and this I would never reach anything which was the first cause of everything else (CSM 78, AT VII 108).

And, in addition to the idea that the *sui causa* is necessary to prove that God exists, it is also necessary to prove his infinity. In fact, Descartes accepts the essential point of Caterus's criticism against Suárez. It is true, he claims, that every limited being is limited by a cause, but this is not self-evident, because someone could think that a being is limited by his own nature. It is therefore useless to employ the principle of the causal dependency to prove God's infinity. On the contrary, he claims:

> What does seem to me self-evident is that whatever exists either derives its existence from a cause or derives its existence from itself as from a cause (CSM II 80, AT VII 112).

Descartes's answer stops here. In the *First Replies* he does not offer any argument showing how this principle, which is self-evident, could prove God's infinity. However, this can be easily explained by the simple fact that Caterus agreed that a being who is positively *a se* is infinite, as we have seen: "If by an act of premeditated choice it were to give itself what it desired, it would undoubtedly give itself all things, and so would be God" (CSM II 68). Insofar as Caterus conceded this point, the only *démarche* that Descartes had to follow was precisely the one that he had actually followed and that constitutes the main concern of the *First Replies*: claiming not to have affirmed that it is impossible that God is the cause of himself and to prove, against Caterus, that the notion of *sui causa* is consistent.

In light of the *First Replies*, the analysis that I have proposed in the second part of this paper, offering a comparison between Descartes's notion of *ens summe perfectum et infinitum* and Suárez's identification between infinite and perfection, should be integrated, both from a conceptual and, in part, historical point of view. However, if, according to Suárez, divine supreme perfection and infinity can be deduced from the fact that God is not caused, according to Descartes, on the contrary, negative aseity does not constitute a sufficient argument for the infinity, that, on the contrary, can be only deduced from positive aseity. But there is much more than this. As is well known, under the pressure of Caterus's objections, Descartes offers in the *First Replies* a new formulation of the proof of God's existence from the Fifth Meditation. In the Fifth Meditation, the demonstration moved from the divine essence, in conformity with the demonstrations of the Mathematics, starting from the essence of mathematical objects to demonstrate their properties (CSM II 45–46, AT VII 66–67). The *First Replies* develops this point, revealing that what is ultimately at work in the application of the mathematical model to God is the notion of a being who "can exist by its own power", that is the *sui causa*: "We shall be unable to think of its existence as possible without also recognizing that it can exist by its own power" (CSM II 85, AT VII 119).

But what is more interesting for our proposal is that, in the formulation of the a priori proof of the *First Replies*, the deduction of God's existence is only the first step of a more general deduction in which, in the second step, all the other perfections of God are deduced from positive aseity:

> And we shall also easily perceive that this supremely powerful being cannot but possess within it *all the other perfections* that are contained in the idea of God (CSM II 85, AT VII 119, mine).

As we have seen, this inference from the positive aseity to the containment of all perfection was already at work in the crucial premise of the proof of the Third Meditation and of its reformulation with Caterus: a being that has the power to give himself his own existence, has also the power to give himself every other perfection.[32] But what at that level was a simple inference, now placed in the heart of the new formulation of the *a priori* proof of God's existence, acquires a quite different meaning. The inference of perfection from aseity that was at the center of the second proof of the Third Meditation, as well as in its reformulation in the *First Replies*, becomes now an a priori demonstration. The *sui causa* becomes the true cause of God's perfections as well as of his existence.

But, in claiming so, Descartes accomplishes precisely the task that Suárez had declared impossible: an a priori demonstration of God's infinity. The inference of divine perfections from positive aseity is indeed the second step of the a priori demonstration of God's existence.

Once again it is impossible to establish whether Descartes knew the text in which Suárez considered and rejected, as we have seen, the possibility of an a priori demonstration of infinity.[33] However, at least one thing can be established as absolutely certain. When, in 1641, Descartes recurs to the notion of a positive aseity to dissolve Caterus's objection against the possibility of deducing the infinity of God from his being *a se*, he was far from recurring to a brand new concept: in Suárez he could find, already formulated, though rejected as impossible, not only the notion of a *sui causa*, but also the idea of an a priori demonstration of God's infinity, that is the idea that the inference from the true, positive essence of God would constitute the only true demonstration of God's infinity.

---

[32] See, *supra*, section 4.

[33] See, *supra*, section 1.

This time it is too difficult to resist postulating an historical relationship between the *First Replies* and the problems revealed by Suárez, once again, in section 2 of his *Metaphysical Disputations* on the attributes of God. It seems that Descartes really wants to play the role of the *voluntarie pertinax* looking for a positive cause of God's infinity who has been blamed in Suárez's *Metaphysical Disputations*.

Abbreviations

AT *Œuvres de Descartes*, éd. par Ch. Adam et P. Tannery, nouv. présent. par J. Beaude, P. Costabel, A. Gabbey et B. Rochot, 11 vols., Paris, Vrin 1964–1974.

CSM *The Philosophical Writings of Descartes*, ed. and transl. by John Cottingham, Robert Stoothoff, Dugald Murdoch, vols. 1 and 2. Cambridge, University Press, 1984–5.

CSMK *The Philosophical Writings of Descartes*, ed. and transl. by John Cottingham, Robert Stoothoff, Dugald Murdoch, and Anthony Kenny, vol. 3: *The Correspondence*, 1991.

References

Agostini, I. (2008). *L'infinità di Dio. Il dibattito da Suárez a Caterus (1597–1641)*. Roma: Editori Riuniti University Press.

Agostini, I. (2010). *L'idea di Dio in Descartes. Dalle Meditationes alle Responsiones*. Firenze-Milano: Le Monnier-Mondadori Education.

Arbib, D. (2017). *Descartes, la métaphysique et l'infini*. Paris: Vrin.

Armogathe, J.-R. (1995). Caterus' objections to God. In R. Ariew & M. Grene (Eds.), *Descartes and his contemporaries: Meditations, objections, and replies* (pp. 34–43). Chicago/London: University of Chicago Press.

Beyssade, J.-M. (1979). *La philosophie première de Descartes: Le temps et la cohérence de la métaphysique*. Paris: Flammarion.

Beyssade, J.-M. (2001). *Descartes au fil de l'ordre*. Paris: PUF.

Birault, H. (1939). *Essai sur les attributs divins chez Descartes. Mémoire pour l'obtention du Diplôme d'études supérieures. Sous la direction de Monsieur Henri Gouhier*. Paris: E.N.S..

Boutroux, É. (1927). *De veritatibus aeternis apud Cartesium. G. Parisiis: Baillière, 1874*. French translation by Georges Canguilhelm. Paris: Alcan.

Carraud, V. (1997). Descartes appartient-il à l'histoire de la métaphysique. In O. Depré & D. Lories (Eds.), *Lire Descartes aujourd'hui* (pp. 153–171). Louvain-la-Neuve/Paris: Institut Supérieur de Philosophie/Peeters.

Chedin, J.-L. (1989). Infini et subjectivité dans la pensée classique. *Revue de Métaphysique et de Morale, XCIV*, 229–250.

Clayton, P. (1992). Descartes and infinite perfection. *Proceedings of the American Catholic Philosophical Association, LXVI*, 137–147.

Clayton, P. (1996a). *Das Gottesproblem. Band 1: Gott und Unendlichkeit in der neuzeitlichen Philosophie*. Paderborn/München/Wien/Zürich: Schöningh.

Clayton, P. (1996b). The theistic argument from infinity in early modern philosophy. *International Philosophical Quarterly, XXXVI*, 5–17.
Coté, A. (2002). *L'infinité divine dans la théologie médiévale (1220–1255)*. Paris: Vrin.
Davenport, A. A. (1999). *Measure of a different greatness. The intensive infinite* (pp. 1250–1650). Leiden/Boston/Köln: Brill.
Devillairs, L. (2004). *Descartes et la connaissance de Dieu*. Paris: Vrin.
Duns Scotus, J. (1639). Quaestiones quodlibetales. In L. Wadding (Ed.), *Opera omnia quae hucusque reperiri potuerunt, collecta, recognita, notis, scholis et commentariis illustrata a P. P. Hibernis Collegii romani S. Isidori professoribus*, 12 voll. (Lugduni: sumptibus Laurentii Durand, 1639), facsimile reprint (Hildesheim: Olms, 1968–1969), vol. XII.
Fabro, C. (1964). *Introduzione all'ateismo moderno*. Roma: Editrice Studium.
Gilson, É. (1932). *L'esprit de la philosophie médiévale*. Paris: Vrin.
Goudriaan, A. (1999). *Philosophische Gotteserkenntnis bei Suárez und Descartesa, im Zusammenhang mit der nederländischen reformierten Theologie und Philosophie des 17*. Brill: Jahrhunderts. Leiden.
Gouhier, H. (1962). *La pensée métaphysique de Descartes*. Paris: Vrin.
Hamelin, O. (1921 [1991]). *Le système de Descartes*. Paris: Alcan.
Koyré, A. (1922). *Essai sur l'idée de Dieu et les preuves de son existence chez Descartes*. Paris: Leroux.
Lazzeroni, V. (1940). *La formazione del pensiero cartesiano e la Scolastica*. Padova: Cedam.
Liard, L. (1903). *Descartes*. Paris: Alcan.
Magnard, P. (1988). La fonction 'Dieu' dans le système. *Revue des Sciences Philosophiques et Théologiques, LXXII*, 195–207.
Marion, J.-L. (1986). *Sur le prisme métaphysique de Descartes*. Paris: PUF.
Maritain, J. (1932). *Le songe de Descartes. Suivi de quelques Essais*. Paris: Buchet/Castel.
Suárez, F. (1597). Metaphysicae disputationes. In *Opera omnia* (pp. 1856–1878, vol. XXVI–XXVII). 28 vol. Parisiis: apud Ludovicum Vivès.
Suárez, F. (n.d.). Tractatus de divina substantia, lib. 2, c. 1, n. 2. In Vivès (Ed.), *Opera omnia* (Vol. I).
Sweeney, L. (1992). *Divine infinity in Greek and Medieval thought*. New York: Peter Lang Publishing.
Vázquez, G. (1598). *Commentaria ac disputationes in primam partem S. Thomae* (2 Vol.). Ingolstadii: Martini Nutti et Ioannis Hertsroy, 1609.

Chapter 3
The Ontic and the Iterative: Descartes on the Infinite and the Indefinite

Anat Schechtman

Abstract Descartes's metaphysics posits a sharp distinction between two types of non-finitude, or unlimitedness: whereas God alone is infinite, numbers, space, and time are indefinite. The distinction has proven difficult to interpret in a way that abides by the textual evidence and conserves the theoretical roles that the distinction plays in Descartes's philosophy—in particular, the important role it plays in the causal proof for God's existence in the *Meditations*. After formulating the interpretive task, I criticize extant interpretations of the distinction. I then propose an alternative at whose core is the idea that whereas the indefinite is a structural, iterative notion, designating the absence of an upper bound, the infinite is an ontic notion, signifying being in general, or what is, without qualification.

3.1 Introduction

In the *Principles of Philosophy*, Descartes makes a surprising claim about infinity: God alone, he claims, is infinite; other entities that might naturally be thought of as infinite, such as numbers, space, and time, are not infinite but indefinite. Descartes's metaphysics thus posits a sharp distinction between the types of non-finitude, or unlimitedness, enjoyed by different entities.

This distinction has long puzzled readers. For, despite the assuredness with which Descartes states it, and despite the fact that it plays an important role in one of his signature arguments in the *Meditations*, the causal proof for God's existence, it has proven difficult to interpret the distinction in a way that neither conflicts with Descartes's other commitments nor inadvertently undermines the distinction itself.

I will begin by examining the chief textual evidence for the distinction, untangling its various components and highlighting several roles that it plays in Descartes's philosophy. I will then discuss ways in which scholars have interpreted the

A. Schechtman (✉)
Philosophy Department, University of Wisconsin-Madison, Madison, WI, USA
e-mail: aschechtman@wisc.edu

O. Nachtomy, R. Winegar (eds.), *Infinity in Early Modern Philosophy*,
The New Synthese Historical Library 76,
https://doi.org/10.1007/978-3-319-94556-9_3

distinction. These interpretations face several difficulties. Identifying their flaws will allow us to formulate a more satisfying interpretation—one that draws on the insights of extant readings yet avoids their problematic consequences. At its core is the idea that whereas the indefinite is a structural, iterative notion, designating the absence of an upper bound, the infinite is an ontic notion, signifying being in general, or what *is*, without qualification.

3.2 Three Distinctions

Let us begin with Descartes's most explicit statement of the distinction between the infinite and the indefinite. It appears in Part I, articles 26 and 27, of the *Principles of Philosophy*.[1] (For ease of reference below, I have divided the passages into segments marked with uppercase letters.)

> 26. [A] We should never enter into arguments about the infinite. Things in which we observe no limits—such as the extension of the world, the division of the parts of matter, the number of the stars, and so on—should instead be regarded as indefinite.
>
> [B] Thus we will never be involved in tiresome arguments about the infinite. For since we are finite, it would be absurd for us to determine anything concerning the infinite; for this would be to attempt to limit it and grasp it. So we shall not bother to reply to those who ask if half an infinite line would itself be infinite, or whether an infinite number is odd or even, and so on. It seems that nobody has any business to think about such matters unless he regards his own mind as infinite. [C] For our part, in the case of anything in which, from some point of view, we are unable to discover a limit, we shall avoid asserting that it is infinite, and instead regard it as indefinite. [D] There is, for example, no imaginable extension which is so great that we cannot understand the possibility of an even greater one; and so we shall describe the size of possible things as indefinite. Again, however many parts a body is divided into, each of the parts can still be understood to be divisible and so we shall hold that quantity is indefinitely divisible. Or again, no matter how great we imagine the number of stars to be, we still think that God could have created even more; and so we will suppose the number of stars to be indefinite. And the same will apply in other cases.
>
> 27. The difference between the indefinite and the infinite
>
> [E] Our reason for using the term 'indefinite' rather than 'infinite' in these cases is, in the first place, so as to reserve the term 'infinite' for God alone. [F] For in the case of God alone, not only do we fail to recognize any limits in any respect [*omni ex parte*], but our understanding positively tells us that there are none. [G] Secondly, in the case of other things, our understanding does not in the same way positively tell us that they lack limits in some respect [*aliqua ex parte*]; we merely acknowledge in a negative way that any limits which they may have cannot be discovered by us. (AT 8A.14-15/CSM 1.201-2)

The distinction introduced in these articles is complex. It is naturally read as having three components, corresponding to three ways in which the infinite and indefinite are meant to be different. We may think of these components as subordinate

[1] References to Descartes's works cite the volume and page number in Descartes (1996) (abbreviated 'AT'), followed by the volume and page number in Descartes (1985–1992), vols. 1 and 2 (abbreviated 'CSM'), or by the page number in vol. 3 (abbreviated 'CSMK'). I use '*Meditations*' for *Meditations on First Philosophy*, and '*Principles*' for *Principles of Philosophy*.

distinctions, of which the distinction between the infinite and the indefinite is composed. The first is a distinction in scope:

> *Scope distinction*: The infinite and the indefinite differ in their extensions. God alone is infinite.[2] God is not indefinite, whereas the extension of the world, the division of the parts of matter, and numbers (e.g., the number of stars), and perhaps other entities, are indefinite. (A, D, and E)

The second is an epistemological distinction:

> *Epistemological distinction*: The infinite and the indefinite differ in how they are perceived by us. We "positively understand" that the infinite is unlimited, whereas we are "unable to discover" limits and hence "merely acknowledge in a negative way" that the indefinite is unlimited. (C, F, and G)

The third is a metaphysical distinction:

> *Metaphysical distinction*: The infinite and the indefinite differ in the way in which each is unlimited (i.e., not limited). Whereas the infinite does not possess "limits in any respect" [*omni ex parte*], indefinite entities "lack limits in some respect" [*aliqua ex parte*] (D, F, and G).[3]

In what follows, I will use one of these labels—'scope,' 'metaphysical,' or 'epistemological'—when discussing one of these three subordinate distinctions ('sub-distinctions', for short). I will speak of *the* distinction when discussing the infinite-indefinite distinction that is composed of the three sub-distinctions.

This taxonomy is not meant to settle substantive questions regarding the correct interpretation of the distinction. The aim is simply to disentangle and highlight the main components of the distinction, as articulated in the passages quoted above. In other words, Descartes's distinction between the infinite and the indefinite—including its epistemological and metaphysical components—remains to be interpreted. We can, therefore, think of the three sub-distinctions as data that any interpretation of the distinction should aspire to uphold, all else being equal.

Another important set of data is provided by the interrelated theoretical roles that the distinction plays in Descartes's philosophy. I will focus on two such roles.

First, as indicated in (B), the distinction is employed by Descartes to avoid "tiresome arguments" provoked by "absurd" queries, such as whether a half of an infinite

[2] In the Fourth Meditation, Descartes says of his own will that it is "so great that the idea of any greater faculty is beyond my grasp; so much so that it is above all in virtue of the will that I understand myself to bear in some way the image and likeness of God." (AT 7.57/CSM 2.40) This has sometimes been read as a statement to the effect that the human will is infinite, perhaps contrary to the claim in the *Principles* that God alone is infinite. (See Naaman-Zauderer (2010, ch. 4) and Boehm (2014) for discussion.) Although I personally do not read Descartes as claiming that the human will is infinite, I will remain neutral on this issue here, for (I think) it can be properly addressed only after the infinite-indefinite distinction has been clarified.

[3] The labels "metaphysical" and "epistemological" are due to Wilson (1999), who speaks of metaphysical and epistemological "criteria". Wilson does not identify a scope distinction (or criterion) as such, though it is implicit in her discussion.

line is itself infinite, or whether an infinite number is odd or even.[4] Whatever the infinite is, it must not invite but rather prohibit these and other, related queries.

Second, and perhaps more significantly, the distinction supports a key premise in Descartes's argument for God's existence in the Third Meditation. The argument consists in two main premises: we have an idea of an infinite being, and given some of the idea's characteristics (specifically, that it has maximum "objective reality"), only an infinite being could have been the cause of this idea. It follows from these two premises that an infinite being exists. Yet it does not follow from these two premises alone that *God* exists—unless, of course, it is assumed that God is the sole infinite being. All three components of the infinite-indefinite distinction contribute to securing this further assumption. The scope distinction asserts that God alone is infinite. The epistemological and metaphysical distinctions jointly underwrite this assertion, by explaining how the unlimitedness of God is different from the unlimitedness of numbers, say, or extension. Furthermore, the epistemological distinction justifies belief in the metaphysical distinction—and, in turn, the scope distinction. For at least part of the reason to think that the infinite and the indefinite are metaphysically distinct, and have different extensions, is that there is a difference in how they are perceived by us. Absent the explanation and justification provided by the epistemological and metaphysical distinctions, the scope distinction's assertion of uniqueness would seem ad hoc or arbitrary, thereby compromising a crucial step in the argument for God's existence.[5]

The distinction may play other roles, in addition to these two. In fact, a third theoretical role will emerge below, when examining the metaphysical distinction.[6] My claim is simply that the two roles identified above are among those that any adequate interpretation must take into account. The interpretive task, as I understand it, is to articulate the distinction between the infinite and the indefinite as clearly as possible, and in a way that not only abides by the textual evidence—in particular, upholds the three sub-distinctions in the *Principles* passage—but also conserves the theoretical roles that the distinction plays in Descartes's philosophy.

---

[4] But cf. Descartes's April 15th 1630 letter to Mersenne (AT 1.146-7/CSMK 23). Descartes's dismissal of certain queries about infinity is not unusual in the period. See Mancosu (1996, esp. chs. 2 and 5) for discussion.

[5] See Schechtman (2014) for further discussion of the argument for God's existence in the Third Meditation. Curley (1978), Ariew (1987), and Wilson (1999) also emphasize the importance of the distinction for the argument.

[6] It has been claimed that the distinction plays a non-theoretical, political role as well, serving to fend off potential Church sanctions that Descartes would face were he to claim that the world is infinite (and consequently, perhaps, not geocentric). Such a role is suggested by Descartes's June 6th, 1647 letter to Chanut: "I recollect that the Cardinal of Cusa and many other Doctors have supposed the world to be infinite without ever being censured by the Church … And my opinion is not so difficult to accept as theirs, because I do not say that the world is *infinite*, but only *indefinite*." (AT 5.51/CSMK 319) Koyré (1957) famously took the distinction to play *only* this political role. But as shown by the theoretical roles discussed in the main text, this position is untenable. For a similar verdict, see Ariew (1987, §3.2). For further discussion of the political role of the distinction, see Vilmer (2011).

The following three sections examine several important ways in which the epistemological and metaphysical components of the distinction have been understood. I will focus on extant interpretations of these components because, as we saw above, they underwrite the scope distinction, which would be ad hoc or arbitrary without them. While each of the interpretations is in some ways natural and suggestive, I will argue that none is fully satisfactory. In the final section, I will develop an alternative interpretation of both components, and of the distinction between the infinite and the indefinite that they compose.

3.3 The Epistemological Distinction: Ignorance

Here, again, is the epistemological component of Descartes's distinction:

> *Epistemological distinction*: The infinite and the indefinite differ in how they are perceived by us. We "positively understand" that the infinite is unlimited, whereas we are "unable to discover" limits and hence "merely acknowledge in a negative way" that the indefinite is unlimited. (C, F, and G)

In this section, I will formulate, and ultimately reject, what I take to be the most prominent interpretation of this distinction.

According to what I will call the *ignorance interpretation* of the epistemological distinction, it refers to the extent of our knowledge, and correlatively, to the extent of our ignorance: Whereas we know that the infinite is unlimited, we do not—indeed, cannot—know that the indefinite is unlimited. Importantly, we also cannot know that the indefinite is limited. We are ignorant about whether it is or is not unlimited, and about whether it is or is not limited.[7]

This interpretation can seem natural in light of Descartes's claim in (G) that "any limits which [the indefinite] may have cannot be discovered by us"—suggesting that, *for all we know, or are capable of knowing*, there may be such limits to discover.[8] If this is correct, then to call something indefinite is simply to declare our ignorance with respect to its limits: for all we know, or are capable of knowing, it may have them, or it may not. Neither possibility is ruled out by what we do or could know.

---

[7] See, e.g., Ariew (1987, 156): "Descartes' indefinite is to be understood as a notion stemming from a defect of our understanding and not from the nature of things." Cp. North (1983) and McGuire (1983). Wilson (1999) offers an interesting and subtle variant on the ignorance interpretation, which I discuss below, in note 11. Janiak (2015) seems to endorse the ignorance interpretation as well, though he acknowledges that it is difficult to reconcile with the existence of the metaphysical distinction, for reasons that will be discussed in the main text shortly.

[8] See also Descartes's February 5th, 1649 letter to More: "God is the only thing I positively understand to be infinite. As to other things like the extension of the world and the number of parts into which matter is divisible, I confess I do not know whether they are absolutely infinite; I merely know that I know no end to them, and so, looking at them from my own point of view, I call them indefinite." (AT 5.274/CSMK 374) Additional passages in this spirit are to be found in the French version of *Principles* 1.27 (AT 9B.37) and the Conversation with Burman (AT 5.154/CSMK 339ff), both of which are discussed by Ariew (1987).

The ignorance interpretation faces several difficulties. First, it undermines rather than justifies belief in any metaphysical distinction between the infinite and the indefinite. For as was just pointed out, on this interpretation, the epistemological distinction is compatible with either of two options: the indefinite is limited, or it is unlimited. But on either option, there is no good reason to think of the infinite and indefinite as involving different types of unlimitedness, as per the metaphysical distinction. For if the indefinite is limited, then of course it involves *no* type of unlimitedness, let alone a type that is different from the one the infinite involves. And if the indefinite is unlimited, for all we know it might enjoy the same type of unlimitedness as the infinite, or God, does.

Apparently assuming the ignorance interpretation, Henry More seems to highlight something like this consequence when he writes in a letter to Descartes from December 11th, 1648:

> I do not comprehend your indefinite extension of the world. For this indefinite extension is either infinite *simpliciter* or infinite only to us. If you mean infinite extension *simpliciter*, why do you hide your meaning with excessively modest words? If you mean infinite only for us, the extension will in reality be finite, for our mind is not the measure of things or of the truth. (AT 5.242; cited in Koyré 1957, 114)

More's reasoning makes explicit the two options identified just above, and concludes that neither delivers the verdict that the indefinite is a distinct metaphysical category. For, he observes, "our mind is not the measure" when it comes to matters such as these: so, the fact that we are ignorant about the status of one entity but not of another does not in itself serve as evidence sufficient to justify the conclusion that the statuses themselves are distinct.[9]

Second, and subsequently, the ignorance interpretation undermines the distinction's ability to play its role in the argument for God's existence in the Third Meditation. As was noted above, the distinction is needed to secure a crucial assumption, namely, that God alone is infinite. It is not enough that God alone is *known by us* to be infinite. But that is all Descartes is entitled to assume, if the ignorance interpretation is correct.

Third, the ignorance interpretation conflicts with other Cartesian texts, where Descartes seems to deny its main contentions. These are passages in which Descartes voices a much stronger commitment, not only to our ignorance about whether indefinite entities are limited or unlimited, but to our inability to even conceive of them as limited. For example, in a letter to Chanut from June 6th, 1647, Descartes writes:

> [I]f we suppose the world to be finite, we are imagining that beyond its bounds there are some spaces which are three-dimensional and so not purely imaginary, as the philosophers'

[9] Proponents of the ignorance interpretation tend to embrace this implication, maintaining that Descartes was unjustified, mistaken, or confused in endorsing the metaphysical distinction. See, e.g., Ariew (1987) and Janiak (2015). See also Leibniz's (1969, 139) remark that "the indefinite of Descartes is not in the thing but in the thinker", and More's verdict in his letter to Anne Conway from May 5th 1651: "For infinite and indefinite in Des Cartes sense, truly Madam, I can not easily absteine from being of your Ladiships opinion in that, that they come much to one" (quoted in Conway 1992, 486–9).

> jargon has it. These spaces contain matter; and this matter cannot be anywhere but in the world, and this shows that the world extends beyond the bounds we had tried to assign to it. Having then no argument to prove, and not even being able to conceive, that the world has bounds, I call it indefinite. (AT 5.51/CSMK 319-20)

Similarly, and shortly thereafter, Descartes writes in a February 5th, 1649 letter to More:

> To remove all difficulties here, I should explain that I call the extension of matter indefinite in the hope that this will prevent anyone imagining a place outside it into which the particles of my vortices might escape, for on my view, wherever such a place may be conceived, there is some matter. When I say that matter is indefinitely extended, I am saying that it extends further than anything a human being can conceive. (AT 5.274-5/CSMK 364)

In these letters, Descartes claims that we cannot conceive of the world as limited. To do this, we would have to imagine the world as bounded in space, and hence as surrounded by empty space. But according to Descartes's theory of space, there is no such thing as empty space: space is always filled with matter, which is itself part of the world. So by trying to conceive of the world as limited, we in fact imagine the world as extending beyond those limits. The exercise of conceiving of the world as limited is self-undermining. A similar exercise can be constructed to show the inconceivability of limits in the case of other indefinite entities.

The significance of these exercises should not be understated. As was explained above, our ignorance as to whether or not the indefinite is limited is to be understood as being compatible both with its being limited and with its being unlimited. But the fact that conceiving of the indefinite as limited is self-undermining reveals that the world is *un*limited, and moreover, that we are in a position to know this. The passages that emphasize our inability to conceive of the limits of the indefinite are therefore in conflict with the ignorance interpretation of the epistemological distinction.

To see why in Descartes's view our inability to conceive of the indefinite as limited is incompatible with its being limited, consider what Descartes says in a second letter to More, from April 15th, 1649:

> It conflicts with my conception, or, what is the same, I think it involves a contradiction, that the world should be finite or bounded; because I cannot but conceive a space beyond whatever bounds you assign to the universe; and on my view such a space is a genuine body. (AT 5.345/CSK 374-5)

Focusing again on the case of the world, Descartes here confirms that our inability to conceive of it as limited entails that it cannot have limits, and is instead unlimited. For our inability to so conceive of it is not due to lack of imagination or insight, but to the fact that what we are trying to conceive involves a *contradiction*. And if what we are trying to conceive involves a contradiction, it follows that it is impossible.[10]

[10] See, e.g., the Sixth Meditation: "I have never judged that something could not be made by [God] except on the grounds that there would be a contradiction in my perceiving it distinctly." (AT 7.71/CSM 2.50) Granted, Descartes famously holds that God is the creator of the eternal truths, and that he could have made necessary truths false, and contradictory claims possible (indeed, true), had he

It might be objected that because the passages just cited are from Descartes's correspondence, they should be granted less weight than passages from his published work, such as (G), on which the ignorance interpretation is based. But, first, (G) can be sensibly understood without the ignorance interpretation (as we will see below).[11] Second, there are other passages from Descartes's published work, including the *Principles*, that seem to conflict with this interpretation in much the same way as those we have been considering. For example, here is article 21 of Part II of the *Principles*:

> 21. Similarly, the extension of the world is indefinite.
> What is more we recognize that this world, that is, the whole universe of corporeal substance, has no limits to its extension. For no matter where we imagine the boundaries to be, there are always some indefinitely extended spaces beyond them, which we not only imagine but also perceive to be imaginable in a true fashion, that is, real. And it follows that these spaces contain corporeal substance which is indefinitely extended. For, as has already been shown very fully, the idea of the extension which we conceive to be in a given space is exactly the same as the idea of corporeal substance. (AT 8A.52/CSM 1.232)

Here Descartes does not say that for all we know, or could know, the world may (or may not) be unlimited. Rather, he explicitly states that the world is unlimited, and that we recognize this to be so. Moreover, the line of reasoning that he invokes in support of this statement is the same one we have seen in the correspondence, proceeding from the inconceivability of the world being limited to the conclusion that it is unlimited.

These considerations provide impetus to consider other interpretations, and to ask whether they are better equipped to fulfill the interpretive task presented above. I submit that on Descartes's view, we know that the infinite is unlimited, and we also know that the indefinite is unlimited. This is of course compatible with the existence of an epistemological distinction, one pertaining to how the unlimitedness of each is known or perceived by us. After all, we "positively understand" that the infinite is unlimited yet merely "acknowledge in a negative way" that the indefinite is unlimited. That we perceive each to be unlimited is also compatible with—though it of course does not entail—a metaphysical distinction, pertaining to the unlimitedness of each, which would explain why God alone is infinite (as required for the Third Meditation argument).

---

chosen to do so. However, as Wilson (1999, 116) observes in a similar dialectical context, it seems clear that Descartes intended for this doctrine to explain, rather than undermine, the necessity of eternal truths, or the impossibility of contradictions.

[11] In her interesting and subtle discussion of the epistemological distinction, Margaret Wilson appears to treat (G) as providing reason to reject the entailment from inconceivability to impossibility in Descartes. She writes (1999, 115): "Descartes's view all along has been, I suggest, that there is *something inconceivable to us* in the idea that the world has limits, some conceptual barrier to positing limits to matter. Yet he seems to hold that this fact does *not* commit him to the view that the world lacks limits." (Earlier in the discussion, she makes it clear that the evidence for the latter claim is (G); *ibid.*, 114). Hence Wilson's interpretation is ultimately a variant on the ignorance interpretation, where ignorance is replaced with inconceivability. However, as suggested in the main text, and as we will see below, (G) can be sensibly understood without appeal to the ignorance interpretation, or to Wilson's variant on it.

In the next two sections, I examine extant attempts to interpret the metaphysical distinction, arguing that they too fail to fulfill the interpretive task we face. Subsequently, I propose an alternative interpretation of the metaphysical distinction, coupled with an (non-ignorance based) interpretation of the epistemological distinction.

3.4 The Metaphysical Distinction: Respects

Let us now turn to the metaphysical component of Descartes's distinction between the infinite and the indefinite:

> *Metaphysical distinction*: The infinite and the indefinite differ in the way in which each is unlimited (i.e., not limited). Whereas the infinite does not possess "limits in any respect" [*omni ex parte*], indefinite entities "lack limits in some respect" [*aliqua ex parte*] (D, F, and G).

According to what I will call the *respects interpretation*, this distinction consists in a difference regarding the number of respects in which the infinite and the indefinite are unlimited: whereas the indefinite is unlimited in some respects, the infinite is unlimited in all respects.[12] Jean-Marie Beyssade seems to have exactly this interpretation in mind when he writes:

> God is positively infinite in all respects and not just with respect to one kind of being [whereas] extended substance…is not [unlimited] in all respects or absolutely infinite. (1979, 313)[13]

The central idea behind this interpretation, as I understand it, can be brought out by considering a geometrical analogy. A geometrical entity can be unlimited in one dimension (i.e., an unlimited line), in two dimensions (i.e., an unlimited plane), or in three dimensions or more (i.e., an unlimited n-dimensional space). Likewise, according to the respects interpretation, non-geometrical entities, such as the number of stars or God himself, can be unlimited in one or more respects.[14]

---

[12]This might be thought to resemble Spinoza's distinction between what is unlimited in its kind, and what is absolutely infinite, or infinite in all kinds. See *Ethics,* part I, definition 6 (Spinoza 1985).

[13]Cp. Kendrick (1998, 31). Vilmer (2008) claims that the majority of writers on the infinite-indefinite distinction adopt a reading along these lines.

[14]The respects interpretation coheres with a certain natural translation of two key elements in the *Principles* passages (quoted in Sect. 3.1), the expressions *omni ex parte* and *alique ex parte* (recall F and G). *Omni* and *alique* are often translated using the quantifiers "all" and "some", while *parte* is translated using the term "respects". It therefore becomes natural to interpret this element as claiming that the infinite does not have limits "in any respect" [*omni ex parte*]. However, this translation is not obligatory; instead, *omni ex parte* can be translated as "completely" or "absolutely", where such completeness need not be understood as holding "in any respect". This option will be exploited by the alternative interpretation of the metaphysical distinction developed in Sect. 3.6.

The respects interpretation offers a plausible perspective on the indefinite. As shown by Descartes's paradigmatic cases of indefinite entities (recall D), they do indeed seem to be unlimited in particular respects. The world is unlimited with respect to extension, numbers are unlimited along the number line, and so on for other instances of the indefinite.

However, the respects interpretation's treatment of the infinite gives rise to decidedly un-Cartesian consequences. For as it stands, there is no restriction on the domain of respects in which God is unlimited. In particular, there is nothing to exclude those respects in which, evidently, God is *not* unlimited. When it is said that the infinite is unlimited in all respects, there is no indication that the quantifier 'all' ranges over a restricted domain—or, if it does, how that domain is restricted. The implication seems to be that God is unlimited in all respects in which something is unlimited. However, one such respect in which something is unlimited is extension. So it follows from this interpretation that God is unlimited in extension. But, of course, Descartes (unlike, say, Spinoza) does *not* think that God is extended, let alone extended without limit.[15]

This problem is not easy to solve. Consider what would be required to modify the respects interpretation so that it does not yield the un-Cartesian consequence just mentioned. What is needed is a way to restrict the domain of respects in regard to which the infinite is unlimited so that it includes all and only respects that are appropriate to attribute to God. The obvious solution is to restrict the domain so that it includes all and only respects that are or constitute *perfections*. In fact, in later work Beyssade appears to do just this. He writes:

> In each class of perfection, for example, knowledge, power, duration, constancy and so on, I have a conception of a more perfect being, and eventually I come to conceive of this perfection as infinite (or, which amounts to the same, as indefinite). Next, I pass in a lateral manner, as it were, from one class of perfection to another, and thus construct the idea of an absolutely infinite, or supremely perfect, being. (Beyssade 1992, 179)

By restricting the domain of respects in this way, this modification of the respects interpretation avoids the un-Cartesian consequence we have just discussed—since by Descartes's lights, at least, being extended is an *im*perfection.[16] But, as I will now argue, it does so at great cost, for it thereby fails to conserve what is arguably the distinction's most important theoretical role.

Recall that the metaphysical distinction helps to explain *why* God alone is infinite, by identifying how the unlimitedness of God is different from the unlimitedness of indefinite entities. This, in turn, enables the scope distinction to secure the crucial assumption in the argument for God's existence in the Third Mediation, without ad hocery or arbitrariness. However, the version of the respects interpretation under consideration threatens to reinstate ad hocery and arbitrariness of the sort that the metaphysical distinction was meant to eliminate. We are now told that the

[15] Cp. Wilson (1999, 115). And contrast Spinoza's *Ethics*, part I, proposition 15, scholium 2.

[16] See, e.g., *Principles* I.23 (AT 8A.13/CSM 1.201): "[T]he nature of body includes divisibility along with extension in space, and since being divisible is an imperfection, it is certain that God is not a body."

infinite and indefinite are metaphysically distinct because the former, but not the latter, is unlimited in all respects that are or constitute perfections—that is, because the former is unlimited *and perfect* whereas the latter is unlimited *and imperfect*. Of course, this is just the difference between God and indefinite things; barring an independent account of perfections (i.e., one that does not make reference to God), the contrast between being unlimited and perfect and being unlimited and imperfect is no more (nor less) than this. But this—the difference between God and indefinite things—just is the scope distinction. Accordingly, on this modification of the respects interpretation, Descartes would be endeavoring to secure the assumption that God alone is infinite by invoking the scope distinction, which in turn is being explained by the metaphysical distinction, which in turn is being explained by reference to the very distinction between God and indefinite things that the scope distinction asserts. That is no explanation at all.

Although I regard this objection as serious, I will not rest my case on it. The reason is that there is another, perhaps deeper problem with the respects interpretation. On this interpretation, no matter how the appropriate respects are understood, the distinction between the infinite and the indefinite does not consist in *different types* of unlimitedness, but only in the *different application* of *one and the same type* of unlimitedness. The infinite differs from the indefinite insofar as it is unlimited in more respects: whereas one being may be unlimited with respect to knowledge, another with respect to power, a third with respect to constancy, and so on, an infinite being is unlimited in *all* of these respects (and more, though perhaps with a restriction to perfections). However, it is not clear that such a difference could justify positing a metaphysical distinction, any more than the difference between the unlimitedness of a line and that of a plane could justify positing such a distinction.[17] Furthermore, such a difference would seem insufficient to underwrite the claim, found in the Third Meditation's argument, that the only viable candidate for the cause of my idea of an infinite being is God himself (i.e., that God is the only being that has the requisite reality to serve as this cause). Such a difference would also render the metaphysical distinction incapable of underwriting the epistemological distinction. The latter distinction makes sense if the infinite and indefinite possess *different types* of unlimitedness. But if they are merely *different applications* of *one and the same type* of unlimitedness, then it is not clear why the way the infinite is known to us should be significantly different from the way the indefinite is known to us. What is known in both cases is *one and the same type* of unlimitedness, which simply applies to a greater or lesser extent.

In short, the worry is that the respects interpretation, if true, would trivialize the metaphysical distinction and thereby not only prevent it from playing its assigned role in the Third Meditation; it would also disallow the distinction from making

[17] At the very least, it is not clear that such a difference justifies the attention it receives in Descartes's publications and correspondence.

another of its contributions, namely, helping to make sense of the epistemological distinction, which otherwise seems brute and mysterious.[18]

3.5 The Metaphysical Distinction: Cardinality

These problems are avoided by interpretations of the metaphysical distinction that posit genuinely different types of unlimitedness. I will call any such interpretation a *types interpretation*. I am convinced that a version of the types interpretation is correct. I also believe, however, that there is a version of this interpretation that, while tempting, is false. I will first present the mistaken version, which I label the *cardinality reading*, and describe the difficulties it faces, before turning in the next section to my preferred version.

The cardinality version of the types interpretation is voiced in the following comments on Descartes's view of God's infinity, by Lawrence Nolan and Alan Nelson:

> Numbers provide a good analogy here. Natural numbers are endlessly augmentable insofar as any specified natural number, no matter how large, has a successor. Descartes would say that this is a kind of potential infinity [i.e., indefiniteness]—we can conceive no limit of natural numbers. But this is not an actual infinity precisely because any particular sequence of natural numbers, no matter how large, is "incomplete" and can always be augmented. What is more, it might be argued that if one understands that natural numbers have no limit, this induces the idea of the cardinality of the natural numbers. Something like this is indicated in the modern mathematical concept (which Descartes would have rejected in this context because of his philosophy of mathematics) of "omega," which is, as it were, the set of natural numbers viewed as complete. In other words, the modern mathematical idea of the cardinality of the natural numbers functions in a way similar to the idea of complete infinity (God) in Descartes's philosophy. (2006, 108)

This passage describes two importantly different types of unlimitedness. The first consists in the absence of an upper bound: for every finite bound or limit, the entity in question exceeds it. In the case of the natural numbers, for example, for any sequence or set of natural numbers, there is another one that goes further—for, as Nolan and Nelson remind us, every natural number has a successor. Similarly, for every finite region or part of extension, there is another that extends further—as shown by the thought experiment invoked in Descartes's letters to Chanut and More (discussed in Sect. 3.3). This type of unlimitedness consists in a structural, iterative property that is identical to or fully grounded in relations between an entity's finite parts: they are such that for each of them, a greater one exists. I will call it *iterative unlimitedness*.

The second type of unlimitedness described by Nolan and Nelson is not identical to or fully grounded in any such relations. It is not simply a matter of an entity's finite parts being such that for each of them, a greater one exists. Rather, the entity is greater than any of its finite parts. This type of unlimitedness consists in a non-

[18] The latter is the distinction's third theoretical role, foreshadowed in Sect. 3.1.

structural, quantitative property of the entity as a whole: the entity has a size or measure, which is greater than the measure of each of its parts.[19] I will call this *quantitative unlimitedness*.[20]

This distinction, between types of unlimitedness, informs the following interpretation of the metaphysical distinction: to be infinite is to be quantitatively unlimited, whereas to be indefinite is to be merely iteratively unlimited.[21] More specifically, and building on an idea explored in the previous section, God is quantitatively unlimited insofar as each and every one of his attributes or perfections, such as knowledge and power, is quantitatively unlimited. The measure of God's knowledge and power (for example) is greater than that of every finitely knowledgeable or powerful being—in much the same way that, as Nolan and Nelson suggest, the cardinality of the set of natural numbers is greater than that of every finite set of numbers. Shortly after the passage just quoted, they write:

> Applying the point about numbers to knowledge, we get the idea of actually infinite knowledge, or omniscience. We might do the same for power and omnipotence, or any other attribute of God. (2006, 108)

Indefinite entities, on the other hand, are not quantitatively unlimited; nor do they possess attributes that are quantitatively unlimited.

One might object that by treating the unlimitedness of God as identical to the unlimitedness of the natural numbers, this cardinality reading flouts Descartes's insistence on the uniqueness of God's unlimitedness, and in particular Descartes's contrast between the infinity of God and the indefiniteness of number. But this objection is too quick. Nowadays we think of the set of natural numbers as having a cardinality, and as quantitatively unlimited in the sense articulated above. But as Nolan and Nelson note, this is arguably not how Descartes thought of the natural numbers. As we have seen, Descartes is very concerned about the implications of treating numbers and other quantities as infinite. Recall that he explicitly shuns the notion of an infinite number (in B). On the cardinality reading, this is because

---

[19] As should be clear, this property is distinct from, and is not implied by, iterative unlimitedness: even if an entity is such that for any one of its parts, a greater one exists, this does not entail that the entity itself is greater than any of its parts. The latter would follow only if the entity as a whole has a measure, and hence is capable of being compared, with respect to measure, to other entities.

[20] The labels 'iterative' and 'quantitative' are from my Schechtman (forthcoming), where I discuss these two kinds of unlimitedness (or two kinds of infinity, as they are called there) in greater detail. I prefer these labels to Nolan and Nelson's. They use "actual" (or "complete" and "perfect") for the infinite, and "potential" (or "incomplete" and "imperfect") for the indefinite. In the Aristotelian tradition, something is potentially infinite only if it is finite, though it is possible for it to become greater without limit. Yet as Nolan and Nelson themselves observe, the collection of natural numbers is not finite: for every natural number there is an actual, and not a merely possible, successor. Similarly, as discussed in Sect. 3.3, Descartes thinks that each region of extension is exceeded by an actual, and not merely possible, greater region. For these reasons, Descartes's indefinite cannot be understood as Aristotelian potential infinity—contra, e.g., Curley (1978, 224), Nolan and Nelson (2006), and Janiak (2015).

[21] I say 'merely' because otherwise the cardinality reading would imply that God is indefinite—the wrong result. Hereafter, I elide 'merely' for ease of exposition.

Descartes denies that the natural numbers are quantitatively unlimited. Instead, he views them as iteratively unlimited: for every natural number, another, greater one exists (recall D). Similarly, the world is iteratively unlimited: for every finite region of the world, a greater one exists (again, recall D). But Descartes does not commit to the world itself having a measure that is greater than that of each of its finite regions. He does not commit to its being quantitatively unlimited.

The cardinality reading seems to contain an important insight about the indefinite, namely, that it can be elucidated through the notion of iterative unlimitedness. However, there are several reasons to think that the cardinality reading incorrectly depicts the infinite when it casts it as quantitative unlimitedness. I will focus on three points.

First, the cardinality reading seems to conflict with the epistemological distinction. As Nolan and Nelson's discussion makes clear, our perception of quantitative unlimitedness is subsequent to, being derived from, our perception of iterative infinity. In other words, the latter engenders the former. As they say, "if one understands that natural numbers have no limit, this induces the idea of the cardinality of the natural numbers."[22] But recall that according to the epistemological distinction, we "positively understand" that the infinite is unlimited, whereas we "merely acknowledge in a negative way" that the indefinite is unlimited. It is hard to see how what we "merely acknowledge in a negative way" can secure positive understanding.

Second, and relatedly, the cardinality reading prevents the distinction from playing its theoretical role in the argument of the Third Meditation. Recall that one of the premises in that argument is that only an infinite being, God, can be the cause of our idea of the infinite. But, as just emphasized, according to the cardinality reading, our idea of the infinite could be secured through our idea of the indefinite.

Third, with respect to the distinction's first theoretical role, the cardinality reading appears to court the kind of absurdities that the distinction was meant, by Descartes, to avoid. The trouble is that the notion of quantitative unlimitedness appears to allow, or even prompt, precisely the sorts of queries that Descartes is concerned to avoid. Just as one might deem it absurd to hold that there can be a quantitatively unlimited line or number, because then we are forced to wonder how long a half of such a line would be, or whether such a number is odd or even, so one might deem it absurd to hold that there can be quantitatively unlimited power or knowledge. Would a being who knows half as many things as God knows have infinite knowledge? Is the number of things God knows odd or even?

---

[22] Elsewhere in their article Nolan and Nelson seem to want to deny this, but their denial rests on two assumptions, both of which are problematic. The first is that the indefinite is equivalent to potential infinity; I explained why this equivalence fails in note 20. The second assumption is that the idea of "that which is subject to augmentation and potentially infinite" cannot be the cause of the idea of "that which is complete and actually infinite" (1996, 110). But the process they describe in the initial quotation in the main text suggests a direct route from the former to the latter.

3.6 The Ontic Reading

I have already indicated my conviction that Descartes's indefinite is to be read as designating iterative unlimitedness. What remains to be found, I think, is a satisfying treatment of the infinite. The respects interpretation maintains that the infinite is unlimited in all respects. The cardinality reading holds that the infinite is quantitatively unlimited. I have voiced objections to both views. I will now propose a third, which not only avoids the difficulties facing extant treatments of the metaphysical distinction, but also paves the way for plausible interpretations of the epistemological and scope distinctions as well.

At various points Descartes posits an intimate connection between divine infinity and being or reality. Such a connection has a long history,[23] and it is endorsed by Descartes on a few occasions, including in the Third Meditation, where he treats the two notions as equivalent.[24] In a commentary on the Third Meditation contained in his April 23rd, 1649 letter to Clerselier, Descartes takes the opportunity to explain this equivalence and what he believes it to imply about our idea of infinity:

> I say that the notion I have of the infinite is in me before that of the finite because, by the mere fact that I conceive being or that which is, without thinking whether it is finite or infinite, what I conceive is infinite being; but in order to conceive a finite being, I have to take away something from this general notion of being, which must accordingly be there first. (AT 5.356/CSMK 377)

Descartes says here that he understands the idea of the infinite as the idea of being *simpliciter*—being in general, or what *is*, without qualification.

To appreciate Descartes's position, it will be helpful to focus on his claim that being comes in degrees: an infinite substance has "more reality" than a finite substance, which in turn has "more reality" than a mode.[25] For Descartes, the diverse degrees of reality possessed by mode (lowest reality), finite substance (intermediate reality), and infinite substance (most reality) are due to these entities' membership in different ontological categories. These categories are not quantitative; they do not indicate differing cardinalities or measures. Rather, they differ insofar as they imply different dependence relations. Modes depend on substances. Finite substances do not depend on modes. But they do depend on the one infinite substance (God), which is itself absolutely independent—it depends on nothing else whatsoever.[26]

---

[23] Descartes's position bears interesting affinities to a traditional view of God in medieval philosophy and theology. This view conceives God as identical to or as possessing unlimited "Being" or "being itself" [*ipsum esse*], and earthly creatures as possessing qualified, limited being derived from God, by whom they were created and on whom they depend. See, e.g., Augustine (1991, 7.10.16).

[24] See especially AT 7.46/CSM 2.31.

[25] This claim is central to the argument for God's existence in the Third Meditation. I will explain how my proposed interpretation handles the crucial assumption in this argument below. Throughout I use 'being' and 'reality' interchangeably.

[26] See the Third Replies (AT 7.185/CSM 2.130). I discuss Descartes's treatment of the relevant dependence relations in Schechtman (2016).

In sum, Descartes views infinity as having the highest degree of reality. The reality of the infinite is unqualified, and equivalent to absolute independence. Such reality is very different from the type of reality we find in the case of the finite, whose reality is qualified: it is a lower degree of reality, equivalent to relative independence.

To possess unqualified reality is to be unlimited in one's being. I will call it *ontic unlimitedness*. My proposal is that Descartes views God's unlimitedness—the unlimitedness of the infinite—as ontic unlimitedness.

This proposal puts us in a position to formulate a version of the types interpretation of the metaphysical distinction that emphasizes, not cardinality, but being or reality. I call it the *ontic reading*. According to this reading, the indefinite is iteratively unlimited; this is the sense in which it "lack limits in some respect" [*aliqua ex parte*]. The infinite, on the other hand, is ontically unlimited; this is the sense in which it is "completely unlimited" [*omni ex parte*].[27]

The ontic reading of the metaphysical distinction also helps to make sense of the epistemological distinction. We "positively understand" that the infinite is unlimited, insofar as we perceive it to possess unqualified reality—that is, to be unlimited in its being (ontically unlimited). By contrast, we "merely acknowledge in a negative way" that the indefinite is unlimited, insofar as we perceive it to be such that it only lacks an upper bound (iteratively unlimited). The fact that what we perceive is a lack, or absence, in the case of the indefinite explains why we "merely acknowledge in a negative way" rather than "positively understand" that what we perceive lacks limits.[28]

I have explained how the ontic reading interprets both the metaphysical and epistemological distinctions. It also delivers an interpretation of the scope distinction: God alone is infinite, since God alone is ontically unlimited. Other unlimited entities are merely iteratively unlimited, and hence indefinite.[29]

Finally, the ontic reading conserves the theoretical roles that the distinction plays in Descartes's philosophy. First, if the ontic reading is correct, then the distinction between the infinite and the indefinite allows Descartes to avoid "absurd" queries such as whether a half of an infinite line is itself infinite. Lines, numbers, and other such entities are iteratively unlimited: for each of their parts, a greater part exists. To say this is *not* to say that they are quantitatively unlimited—that they have a measure and can be compared, measure-wise, to their parts. Hence, the assertion that they are indefinite does not give rise to the question of whether, for example, half of an infinite line is itself infinite. At the same time, the infinite is also not quantitatively unlimited; it is ontically unlimited. Hence, analogous, "absurd" queries do

[27] Recall note 14, where I noted that *omni ex parte* can be translated as "completely" or "absolutely".

[28] In this way, all of (C), (F), and (G) can be sensibly interpreted without appealing to our ignorance.

[29] This serves to elucidate (A), (D), and (E).

not arise in the case of God either (as they did on the cardinality reading): there is no occasion to ask, e.g., whether the number of things God knows is odd or even.[30]

Second, the ontic reading fits well with the distinction's role in securing the crucial assumption in the Third Meditation argument, namely, that God is the sole infinite being. First, that is what is asserted by the scope distinction, on this reading. Second, the reading implies that the epistemological and metaphysical distinctions jointly underwrite this assertion, by explaining how God's unlimitedness differs from the unlimitedness of indefinite entities: to wit, the former is ontic whereas the latter is iterative. Furthermore, on the ontic reading, the epistemological distinction is able to justify belief in the metaphysical distinction. For the fact that in one case we perceive something as having unqualified reality, whereas in the other case we only perceive something as lacking an upper bound, provides evidence that there are two ways in which things are unlimited. It follows that, on the ontic reading, the crucial assumption is not at risk of seeming ad hoc or arbitrary (as it was on the respects interpretation).

I suggested at the outset that the interpretive task is to articulate the distinction between the infinite and the indefinite as clearly as possible, and in a way that both abides by the textual evidence and conserves the theoretical roles that the distinction plays in Descartes's philosophy. I submit that the interpretation I have developed here fulfills this task—and, moreover, that it does so in a way that highlights the importance of ontic categories (reality, being, and independence) to Descartes's theory of the infinite.[31]

Abbreviations

AT	Descartes, René. *Oeuvres de Descartes*. Edited by C. Adam and P. Tannery. 12 vols. Paris: J. Vrin, 1996.
CSM	*The Philosophical Writings of Descartes*. Translated by J. Cottingham, R. Stoothoff, D. Murdoch. 2 vols. Cambridge: Cambridge University Press, 1984–5.
CSMK	The Philosophical Writings of Descartes, ed. and transl. by J. Cottingham, R. Stoothoff, D. Murdoch, and A. Kenny, vol. 3: *The Correspondence*, 1991.

[30] This upholds (B).

[31] I have received helpful suggestions from audience members at a conference on infinity in early modern philosophy at the Van Leer Institute in Jerusalem and at the 2017 Atlantic Canada Seminar in Early Modern Philosophy at Dalhousie University. I am also grateful to Ohad Nachtomy for helpful comments, and in particular to John Bengson for extensive input, both critical and constructive.

References

Ariew, R. (1987). The infinite in Descartes' conversation with Burman. *Archiv für Geschichte der Philosophie, 69*(2), 140–163.

Augustine (1991). *Confessions* (C. Henry, Trans.). Oxford: Oxford University Press.

Beyssade, J. M. (1979). *La philosophie première de Descartes*. Paris: Flammarion.

Beyssade, J. M. (1992). The idea of God and the proofs of his existence. In J. Cottingham (Ed.), *The Cambridge companion to Descartes* (pp. 174–199). Cambridge: Cambridge University Press.

Boehm, O. (2014). Freedom and the Cogito. *British Journal for the History of Philosophy, 22*(4), 704–724.

Conway, A. (1992). In M. Nicolson & S. Hutton (Eds.), *The Conway letters: The correspondence of Anne, Viscountess Conway, Henry more, and their friends* (pp. 1642–1684). Oxford: Oxford University Press.

Curley, E. (1978). *Descartes against the skeptics*. Cambridge: Harvard University Press.

Janiak, A. (2015). Mathematics and infinity in Descartes and Newton. In V. De Risi (Ed.), *Mathematizing space* (pp. 209–230). Basel: Birkhäuser.

Kendrick, N. (1998). Uniqueness in Descartes' 'infinite' and 'indefinite'. *History of Philosophy Quarterly, 15*(1), 23–36.

Koyré, A. (1957). *From the closed world to the infinite Universe*. Baltimore: Johns Hopkins University Press.

Leibniz, G. W. (1969). In L. E. Loemker (Ed.), *Philosophical papers and letters*. Dordrecht: Reidel.

Mancosu, P. (1996). *Philosophy of mathematics and mathematical practice in the seventeenth century*. New York: Oxford University Press.

McGuire, J. E. (1983). Space, geometrical extension, and infinity: Newton and Descartes on extension. In W. R. Shea (Ed.), *Nature mathematized* (pp. 69–112). Reidel: Dordrecht.

Naaman-Zauderer, N. (2010). *Descartes' deontological turn*. Cambridge: Cambridge University Press.

Nolan, L., & Nelson, A. (2006). Proofs for the existence of God. In S. Gaukroger (Ed.), *The blackwell guide to Descartes' meditations* (pp. 104–121). Oxford: Blackwell.

North, J. (1983). Finite and otherwise. Aristotle and some seventeenth century views. In W. R. Shea (Ed.), *Nature mathematized* (pp. 113–148). Reidel: Dordrecht.

Schechtman, A. (2014). Descartes's argument for the existence of the idea of an infinite being. *Journal of the History of Philosophy, 52*(3), 487–517.

Schechtman, A. (2016). Substance and independence in Descartes. *Philosophical Review, 125*(2), 155–204.

Schechtman, A. (forthcoming). Three infinities in early modern philosophy. *Mind*.

Spinoza, B. 1985. *The collected works of Spinoza*. Vol. 1 (Edwin Curley, Ed. and Trans.). Princeton: Princeton University Press

Vilmer, J. B. J. (2008). La veritable nature de l'indéfini cartésien. *Revue de Métaphysique et de Morale, 60*, 503–515.

Vilmer, J. B. J. (2011). L'indéfini cartésien entre politique et langage. *Revue philosophique de Louvain, 109*(3), 443–460.

Wilson, M. (1999). Can I be the cause of my idea of the world? (Descartes on the infinite and the indefinite). In M. Wilson (Ed.), *Ideas and mechanism: Essays in early modern philosophy* (pp. 108–125). Princeton: Princeton University Press.

Chapter 4
Descartes on the Infinity of Space vs. Time

Geoffrey Gorham

Abstract In two rarely discussed passages – from unpublished notes on the *Principles of Philosophy* and a 1647 letter to Chanut – Descartes argues that the question of the infinite (or indefinite) extension of space is importantly different from the infinity of time. In both passages, he is anxious to block the application of his well-known argument for the indefinite extension of space to time, in order to avoid the theologically problematic implication that the world has no beginning. Descartes concedes that we always imagine an earlier time in which God might have created the world if he had wanted, but insists that this imaginary earlier existence of the world is not connected to its actual duration in the way that the indefinite extension of space is connected to the actual extension of the world. This paper considers whether Descartes's metaphysics can sustain this asymmetrical attitude towards infinite space vs. time. I first consider Descartes's relation to the 'imaginary' space/time tradition that extended from the late scholastics through Gassendi and More. I next examine carefully Descartes's main argument for the indefinite extension of space and explain why it does not apply to time. Most crucially, since duration is merely conceptually distinct from enduring substance, the end or beginning of the world entails the end or beginning of real (as opposed to imaginary or abstract) time. In contrast, extension does not depend on any enduring substance besides itself.

4.1 Introduction

Although one prominent commentator has declared that the infinite plays so important a part in the philosophy of Descartes that "Cartesianism may be considered as being wholly based on that idea" Koyré (1957, 106), Descartes himself warns us that "we should never enter into arguments about the infinite" (AT 8A 14; CSM 1

G. Gorham (✉)
Philosophy Department, Macalester College, Saint Paul, MN, USA
e-mail: ggorham@macalester.edu

O. Nachtomy, R. Winegar (eds.), *Infinity in Early Modern Philosophy*,
The New Synthese Historical Library 76,
https://doi.org/10.1007/978-3-319-94556-9_4

201). Descartes hoped to forestall "tiresome arguments" about the infinite by substituting the more modest, anthropocentric notion of the indefinite: "in the case of anything in which, from some point of view, we are unable to discover a limit, we shall avoid asserting that it is infinite and instead regard it as indefinite" (AT 8A 15; CSM 1 202). But although familiar conceptual paradoxes about the infinite – such as whether the infinite is odd or even – are perhaps sidestepped in this way, long-standing cosmological problems about the extent and duration of the world are merely re-formulated: from our point of view, can we discern whether the world has limits in space and time?

Descartes maintains that the world is spatially indefinite but temporally limited. He therefore departs from the standard seventeenth-century method of treating space and time as analogous, with conclusions about the structure of space typically extended, with minimal adjustments, to time.[1] For Descartes, the indefinite extension of space (and hence body) was a conceptual necessity (notwithstanding the incomprehensible power of God). In the *Principles of Philosophy* (II, 21) he insists that wherever we imagine boundaries to extension we are forced to recognize that there must be indefinitely extended space beyond them "which we not only imagine but also perceive to be truly imaginable, that is, real" (AT 8A 52; CSM 1 232). Similarly, he explains to Henry More: "I think it involves a contradiction that the world should be finite; because I cannot but conceive a space beyond any bounds you assign to the universe; and on my view such a space is a genuine body" (AT 5 345: CSMK 375). Put simply, Descartes's argument is that we find it inconceivable that extension (hence body; hence the world) has limits.

It might seem that we find it just as difficult to conceive temporal limits on the world. That is, we find it hard to conceive how a motion, a body, or the world as a whole, could begin (or end) without being preceded (or followed) by time. If so, it should follow for Descartes that duration is indefinite in both temporal directions just as extension is indefinite in all spatial directions. However, although he does not raise this issue in either published versions of the *Principles*, it evidently occurred to Descartes that his reasoning about indefinite extension in *Principles II*, 21, mentioned above, would seem to make extension and duration equally indefinite. For in his own annotations to the *Principles* he notes, and attempts to allay, the "fear that in philosophizing about the indefinite extension of the World we should find its duration likewise mounting to infinity" (AT 11 656).[2] He observes, for one thing, that faith precludes an eternal world (at least looking backwards). Thus, in the annotation Descartes references his own admonition at the end of Part One of the *Principles*, that "divine authority must be put before our own perception" (8A 39; CSM 1 221). But he also declares it certain that our "natural reason" (*ratione naturali*) cannot decide for us the question of the world's beginning. Unfortunately,

[1] See further Gorham (2012).

[2] The annotations were found among papers owned by Leibniz, who titled them 'Annotations which Descartes seems to have made on his *Principles of Philosophy*' (AT 11 545). The date is unknown, though obviously between 1644 and 1650. On their origin and authenticity, see AT 10 207-210; AT 11 647a and 654a. Judging by the surrounding notes, it is plausible that Descartes is here commenting on *Principles* II, 21.

his annotation leaves the matter there, deferring to faith. I think it is worth inquiring why Cartesian natural reason – which is so confident about the indefinite extension of the world – is incapable of discerning its duration (whether indefinite or not). The answer I will offer depends on a fundamental ontological asymmetry between Cartesian space and duration.

4.2 Letter to Chanut I: Space

Fortunately, Descartes takes up the issue in more detail in a 1647 letter to Chanut, who had passed on Queen Christina's concerns that Descartes might be among those who hold the world to be infinite, and that such a doctrine was injurious to the Christian religion (May 11, 1647; AT 5 21). In response, Descartes first puts himself in the good company of the "Cardinal [Nicholas] of Cusa," who "supposed the world to be infinite without ever being censured by the Church" (June 6, 1647; AT 5 51; CSMK 319). And he points out that we honor God to the extent we represent his works as great. He goes on to emphasize, as in the *Principles* and annotations, that he maintains only that the world is *indefinite* rather than *infinite*. In the *Principles* themselves (I, 26-7) he explains that the term 'infinite' should be reserved for God, the sole being we understand positively to have no limits. We should say a being is indefinite when we know only negatively that "we are unable to discover a limit" (AT 8A 15; CSM 1 202. See also AT 5 356; CSMK 377).

This account seems to conflate several distinct senses in which we might be unable to discover a limit:

(i) any limit we discover could conceivably be greater.
(ii) we cannot discover any limit.
(iii) we cannot conceive (or find contradictory) any limit.[3]

Consider the examples of undiscoverable limits Descartes gives in the *Principles*: the number of stars; the divisibility of quantity; the extension of the universe. The number of stars seems to be indefinite in the weakest sense above (i). He says that however many stars we imagine there to be, "God could have created more" (ibid.). But this seems too weak a sense of indefinite for Descartes's conception of space, since it would include even things with limits we might actually determine, such as the number of humans or the size of the earth. (In fact, Descartes drops the 'number of stars' example from all subsequent discussions of the indefinite.) The stars might also qualify under (ii) since it may be that they are too numerous or remote for

[3] Wilson notices this "haziness" in the characterization of indefiniteness but offers a slightly different taxonomy: (i) we notice no limit; (ii) we cannot conceive a limit; (iii) a limit is repugnant or contradictory (1999, 113). I am unsure about her distinction between (ii) and (iii) since Descartes says to More: "It conflicts with my conceptions or, what is the same, involves a contradiction that the world should be finite or bounded; because I cannot but conceive a space beyond whatever bounds you assign to the universe" (AT 5 345; CSMK 374-5). For recent, illuminating discussions of the infinite/indefinite distinction in Descartes see Ariew (1987), Kendrick (1998) and Schectman (this volume).

beings like us to gauge even though we can easily conceive that there is a definite limit. (I'll later suggest that the duration of the universe is indefinite in at most the senses of (i) or (ii).) The divisibility of the quantity and extent of the universe seem to be indefinite in the strongest sense of (iii) and it is this sense that is most important to Descartes. Thus when it comes to his specific arguments about indefinite divisibility and extent later in the *Principles*, he emphasizes that we are unable to conceive a genuine limit to the division and extent of matter.[4]

(iii) is also the crucial sense of 'indefinite' in all the passages that address indefinite spatial vs. temporal extent. Thus, returning to Descartes's response to Queen Christina, he admits that although we lack a positive reason to know the world is infinite, can we neither prove nor even conceive that it is finite. For "if we suppose the world to be finite we are imagining that beyond its boundaries there are some spaces which have three dimensions of their own and so are not purely imaginary, as the philosophers label them" (AT 5 52; CSMK 319). He once again reassures the Queen (via Chanut) that he does not assert that the world is absolutely infinite since God may know it to be limited for reasons incomprehensible to us; nevertheless, since he finds himself unable to conceive that the world has bounds he concludes that the world is indefinite. Descartes then turns to the question whether the same logic proves the duration of the world to be indefinite (in the future and past). Before examining his treatment of this question, it will prove worthwhile to digress briefly on the historical context of Descartes's analysis.

4.3 Digression: Imaginary Space

In an important, late defense of the indefinite extension of the world, Descartes explicitly contrasts his own view with those "who call this space imaginary and thus regard the world as finite" (AT 5 345; CSMK 375). In both this letter to Henry More, and in the letter to Chanut, Descartes is adverting to a long tradition within late scholasticism analyzing the ontological status of the space we 'imagine' beyond the limits of our finite world.[5] The metaphysics of Suárez, for example, which was known to Descartes,[6] admits an 'imaginary space' beyond the finite realm of the cosmos in order to conceive the place of the outermost sphere, the power of God to translate the entire world, and so on. And imaginary time or succession is likewise permitted for parallel reasons. However, Suárez is at pains to emphasize that imaginary space and time are mere conceptual tools or 'beings of reason' in contrast with the real space and time of bodies and motion: "We conceive of this imaginary space as having dimensions. But so conceived this space is a mere being of reason, a

[4] AT 3 477; CSMK 3 202; AT 5 273; CSMK 363.

[5] The definitive treatments of this tradition are Duhem (1987) and Grant (1981). Des Chene (1996) also provides an insightful discussion, giving particular attention to Descartes. On imaginary time, especially in Suárez, see Daniel (1981) and Bexley (2012).

[6] AT 7 235; CSM 2 164

negation or privation . . . this is also true in the example of imaginary succession, which we conceive apart from real time" (DM 54, IV, 7). Among non-Aristotelian contemporaries of Descartes, Gassendi also relied on the imaginary space/time tradition in defense of an infinite universe: "by the words 'space' and 'spatial dimensions' we do not mean anything but that space which is commonly called 'imaginary' and which the majority of sacred doctors admit exists beyond the world" (Brush 389). Gassendi is more guarded than Descartes about the ontology of space, emphasizing with the scholastics that it is *nihil positivam* since it falls outside the substance/accident dichotomy. Nevertheless, he insists that "space and time must be considered real things, or actual entities, for although they are not the same sort of thing as substance and accident they still exist and do not depend on the mind" (Brush 384-5). Both extramundane space and premundane time are real: "the universe could have been created a thousand years before the creation . . . because time then flowed, of which the revolutions of the sun such as we now have could have been an adequate measure" (Brush 397).[7] So although Suárez and Gassendi draw different conclusions about the ontological status of space and time, they both assume that what goes for the one, goes for the other.

4.4 Letter to Chanut II: Time

Returning to Descartes we find, in contrast, that he resists the implication that imaginary time is as real and indefinite as imaginary space. In the letter to Chanut, after rehearsing his argument for the indefinite extension of the world, he turns to the worry (not raised by Christina) that this proves too much, namely that the world also has no beginning or end in time. Descartes remarks that comparing the extension of the world with its duration in this respect merely occasions the thought "that there is no imaginable time before the creation of the world, in which God could not have created it if he had so willed" (AT 5 52; CSMK 320). His point does not seem to be that we cannot imagine time before creation, since we clearly can, but simply that God could have created the world at any such imagined earlier time. Of course, there is likewise no imaginable space beyond the boundaries of a putatively finite world in which God could not have created matter. The difference is that natural reason alone tells us that God really did create extension/matter in the indefinite space we imagine; but in the case of time Descartes insists that the always earlier imaginable duration of the world does not entail that God "really did create it an indefinitely long time ago" (AT 5 53; CSMK 320). He argues as follows:

> For the actual or real existence of the world during the last five or six thousand years is not connected to the possible or imaginary existence which it might have had before then in the way the actual spaces one conceives surrounding a globe (i.e. the world as supposed infinite) are connected with the actual existence of the same globe (ibid.).

The alleged dis-analogy in the 'connections' between imaginary and actual spaces vs. imaginary and actual duration is not immediately evident. Perhaps

[7] See LoLordo (1997).

Descartes's point is simply that, given the conceptual interdependence of space and body, imaginary space entails actual body; but there is no parallel interdependence between time and body. But this would not show that we can conceive a beginning or end to time itself, on Descartes's terms. For his proof of the indefinite extension of the world would still imply the indefinite extension of mere space even if he conceded the possibility of a finite world surrounded by an extramundane vacuum. This is most evident in the *Principles* version of the proof, which proceeds in two steps. He first maintains that the space we inevitably imagine beyond any putative limits to the world is not something we "merely imagine" but something "imaginable in a true fashion, i.e. real (*vere imaginabilia, hoc est realia esse*)" (AT 8A 52; CSM 1 232). The next step is that "it follows that these spaces contain corporeal substance which is indefinitely extended" (Ibid.). This follows from the reality of the space beyond limits, he points out, given the identity of space and body established a few sections earlier (II, 11). So even someone who rejected this second step, and insisted on a real distinction between extension and body, would have to admit an indefinite extramundane void based on the first step. Descartes himself makes this point in response to More, who invoked the famous Archytas thought experiment (AT 5 312) of extending a sword at the edge of a supposedly finite world: "When you imagine a sword going through the limits of the universe, you show that you too do not conceive the world as finite. . . though you give the name 'vacuum' to what you conceive" (AT 5 345; CSMK 375). Similarly, the fact that we imagine always earlier (later) times would entail an indefinite past (future), according to Descartes's argument, even if that past (future) is empty rather than full of matter and motion.

4.5 Cartesian Duration and Time

But perhaps this is enough for Descartes's purposes: he can allow indefinite empty time before the creation of the world so long as the religiously problematic indefinite pre-existence of *the world* is blocked. The problem is that Descartes's metaphysical principles seem to preclude empty time no less than empty space. To see why, we need to examine briefly his metaphysics of time.[8] Contrary to the long Aristotelian tradition that made time 'the number of motion', Descartes insists that "the duration which we find to be involved in movement is certainly no different from the duration involved in things which do not move" (AT 8A 27; CSM 1 212). It is true that in order to measure this duration common to all things, "we compare their duration with the greatest and most regular motions, which give rise to years and days, and call this duration 'time'" (Ibid.). But Descartes insists that we must not conflate duration itself, which is intrinsic to all things, with its temporal measure, which is an intellectual abstraction: "when time (*tempus*) is distinguished from duration taken in the general sense (*duratione generaliter*) and called the measure of movement (*numerum motum*), it is simply a mode of thought" (Ibid.).

[8] For a more detailed discussion, see Gorham (2007).

But what is duration itself? Several sections earlier in the *Principles*, Descartes lists duration (along with substance, order and number) among the most general categories which "extend to all classes of things" whether thinking or extended (AT 8A 23; CSM 1 208). A little later he says we can have a distinct understanding of these universal or transcendental categories only if we do not assign to them the status of substance. Thus, rather than being anything separate from the thing which endures "we should regard the duration of a thing as simply a mode under which we conceive the thing insofar as it continues to exist" (AT 8A 26; CSM 1 211). His point here is not that duration is a mere mental abstraction, or mode of thought, like time. For he goes on to classify duration as an unchanging attribute: "that which always remains unmodified – for example existence or duration in the thing which exists or endures – should be called not a quality or mode but an attribute" (AT 8A 26; CSM 1 211-2). As such, duration and existence are not mere modes of thought but rather ways in which anything can be conceived. Indeed, he observes that "since a substance cannot cease to endure without also ceasing to be, the distinction between a substance and its duration is merely a conceptual one" (AT 8A 39; CSM 1 214). In the French edition of the *Principles*, he observes that all such attributes "allow us to have different thoughts about a single thing" (AT 9B 53).

We can now see why Descartes can allow neither time nor duration before/after creation: not time, because it is a mere mode of thought, and not duration, because it cannot be really separated from the continuance in existence of things. He makes this explicit in the correspondence with More. More suggested: "If God annihilated the universe and created another one out of nothing much later, this 'between-world' or 'world-absence' would have its own duration whose measure would be days, years and centuries. There is therefore a duration of something that does not exist" (AT 5 302). Descartes's reply is blunt: "it involves a contradiction to conceive of any duration intervening between the destruction of an earlier world and the creation of a new one" (AT 5 343; CSMK 373). This response confirms my point that Descartes cannot avoid a world indefinitely extended in time, any more than a world indefinitely extended in space, by separating time or duration from the world.

His dismissal of 'between-world' duration also reveals how similar Descartes's reasons for rejecting empty time are to his reasons for rejecting empty space. Extension and duration are both attributes, the latter general or transcendental and the former peculiar to bodies. But we have already noted that attributes are merely conceptually distinct from their substances. So to conceive an extension distinct from corporeal substance is to conceive an extended nothing. Thus, Descartes rejects extension without body because "it is a complete contradiction that a particular extension should belong to nothing" (AT 8A 49; CSM 1 230).[9] We can now see that he rejects More's speculation about a duration of "something that does not exist" as contradictory for the same reason. It invites us to conceive duration apart from any continuation in existence. But for time to pass between worlds, with nothing intervening, is no less contradictory than for the sides of a vessel to remain apart though nothing is between them.

[9] See also AT 8A 50; CSM 1 231; AT 5 223; CSMK 358.

4.6 Creation from Eternity

So Descartes cannot allow duration to extend indefinitely backward without admitting an eternal world, anymore than he can allow space to extend indefinitely beyond a limited material cosmos. Apart from falling back on faith, one way out of the problem would be to demonstrate from natural reason that the world could not have existed from eternity. But on this famous question, Descartes seems to adopt a quite liberal position. In conversation with Burman, for example, he reportedly declares: "I do not see why God should not have been able to create something from eternity" (AT 5 155; CB 15). The topic at this point in the conversation is the second of the Third Meditation's proofs of God's existence, which establishes that I would need to be continuously created by God even if "I have always existed as I do now" (AT 7 78; CSM 2 33). Indeed Descartes seems committed to eternal creation by his well-known doctrine of the divine origin of the eternal truths: "from all eternity he willed and understood them to be" (AT 1 152; CSMK 25). Descartes also makes clear that an infinite or indefinite past is not absurd or contradictory in itself. When Burman pushes him on the suggestion that God might have created the world from eternity – "but then there would have been an infinite number" (AT 5 155; CB 16) – Descartes denies there is any absurdity in this. He points out that there is no less an infinite division within any given finite quantity. Indeed, Descartes seems to think the latter sort of infinite division within matter is *required* under certain physical conditions (AT 8A 59-60; CSM 1, 239).[10] He observes furthermore that we believe, as a matter of faith, that there is an infinite number with respect to the future, "so why shouldn't it be the same with respect to the past?" (AT 5 155; CB 16).

4.7 The Mereological Independence Doctrine

So far, the asymmetry remains unexplained: Descartes should say – faith notwithstanding – that the world is indefinite in extension *and* duration. But let us return once again to the detailed letter to Chanut and consider more carefully Descartes's insistence that the possible earlier existence of the world is not "necessarily connected (*necessairement jointe*)" to its actual later existence in the strong way that real extension is implicated beyond any imaginable boundary of the world. The disanalogy in these connections does not seem to arise from differences in our conceptions. For Descartes acknowledges that we find it difficult to conceive a temporal beginning, though his expression is somewhat tortured: "*il n'y a point de temps imaginable avant la creation du monde quel Dieu n'eust pu le creer, s'il eust voulu*" (AT 5 53; CSMK 320). He does not explicitly say that we positively imagine earlier times, only that there is no imaginable time unavailable for earlier creation. But his argument doesn't seem to turn on such cautious phrasing. For he goes on to say

[10] See also AT 4 113; CSMK 232; AT 5 274; CSMK 364; AT 7 113; CSM 2 81.

explicitly that the world's "possible or imaginary existence" earlier does not entail its actual indefinite existence. So we do positively imagine earlier times, just like space beyond the world; but for some reason only the latter carries ontological import.

Descartes does not reveal the reason until the very terse conclusion of his argument. After asserting the crucial dis-analogy in the 'connections' between actual and imagined space vs. time, he turns to a seemingly different issue: the eternal future duration of the world (which faith teaches). He says the indefinite future provides a more solid reason to infer its indefinite past duration than does its indefinite extension. But the former is not a solid reason at all: "no one infers" eternal creation in the past from the promise of eternal life (AT 5 53; CSMK 320). The relevance of this apparent red herring to the crucial dis-analogy with extension is made clear only in the final clause of the paragraph, where he states that we cannot infer the eternity of the world "because every moment of its duration is independent of every other" (Ibid.).This, of course, is the much-discussed premise of the second Third Meditation proof of God's existence. In that proof, it is first restricted to the meditator's own duration: "a lifespan can be divided into countless parts each completely independent of the other" (AT 7 49: CSM 2 48). But Descartes later extends it to bodies as well: "The separate divisions of time do not depend on one another. Hence the fact that the body in question is supposed to have existed up until now 'from itself,' that is, without a cause, is not sufficient to make it continue to exist in the future" (AT 7 110; CSM 2 110).[11] From this he derives his version of the doctrine of continuous creation – the entire universe must be "recreated" by God at each instant so that "the distinction between preservation and creation is only a conceptual one (AT 7 49; CSMK 33) – which not only requires God's existence but also grounds the laws of nature in the *Principles of Philosophy* (AT 8A 61; CSM 1 240).

Besides these two functions – proving God and the laws of nature – we can now see that Descartes has a third role for the versatile doctrine of the independence of temporal parts: blocking a seemingly Cartesian argument from natural reason for the indefinite duration of the world. Granted that I always may, perhaps irresistibly, imagine an earlier/later time, it does not follow that what I imagine is real. For the temporal parts of any enduring thing are not conceptually connected in such a way that bordering parts of time guarantee one another's existence: "I regard the divisions of time as being separable from each other so that the fact that I exist now does not imply that I shall continue to exist in a little while" (AT 7 109; CSM 2 78-9). While I can imagine that I go on living indefinitely, it is entirely up to God whether this hope is realized. But the situation is different with extension, I would like to suggest. When we consider the plenum abstractly, as geometrical *res extensa*, there is a sense in which its parts are mutually dependent. We can see this if we suppose any part of extension persisting while a part next to it is destroyed. Either a vacuum would then exist, which is impossible on Cartesian principles, or some other part of

[11] Versions of the argument and doctrine are repeated frequently. See AT 6 35, 45; AT 8A,13; AT 7 109, 165, 369–370; AT 5 45, 53, 155; CSM 1 128–129, 133, 200; CSM 2 78–79, 116, 254–255; CSMK 320; CB, 15–16. See Gorham (2004).

extension would 'move in' to fill the impossible gap (AT 5 272-3; CSMK 363). This explains, I believe, why Descartes says in the Synopsis to the *Meditations* that while individual bodies can perish since they are merely a collection of accidents, "body in general (*corpus in genere*) can never perish" (unless God withdraws his concurrence) (AT 7 10; CSM 2 14). This mutual interdependence of parts also explains why extension or body in general is indefinite.[12]

This does not mean that extension, unlike duration, is independent of God.[13] But I think it does mean that if God creates any part of *res extensa* then the entire indefinite plenum is necessitated as well, in one fell swoop. This is suggested in the creation story of *Le Monde*, which again exploits the imaginary space tradition: "a whole new [world] which I shall bring into being before your mind in imaginary spaces" (AT 11 31; CSM 1 90). Pre-existing imaginary space is, as it were, presupposed. While acknowledging that "the philosophers tell us that such spaces are infinite", in order to prevent confusion Descartes asks us to confine our imagination "to a determinate space which is no greater, say, than the distance between the earth and the principal stars in the heavens" (AT 11 32; CSM 1 90). It is clear that he is not supposing that God might create a finite world, for he insists that "the matter which God has created extends indefinitely far in all directions" (Ibid.). And we should think of this indefinite matter as a "real perfectly solid body which uniformly fills the entire length, breadth and depth of this vast space in the midst of which we have brought our mind to rest" (AT 11 33; CSM 1 91). The "idea of this matter is included to such an extent in all the ideas that our imagination can form that you must necessarily conceive it or you never imagine anything at all" (AT 11 35; CSM 1 92). Indefinite space as a whole is thus implied by the idea of any finite extended thing.

A similar point is indicated in a 1640 letter to Regius, who commented on a draft of the *Meditations*. Descartes maintains that I could not have the positive of idea of God's infinite perfections "unless we derived our origin from a being in which they

---

[12] This, of course, is a version of the argument Spinoza would later champion that extended substance is not really divisible: "Since there can be no vacuum in nature (as I discuss elsewhere) but all their parts must concur so there is no vacuum, it follows that they cannot be really distinguished" (Ethics I P15 Schol (iv); C 96). There are many excellent discussions of Descartes's theory of corporeal substance, such as Slowik (2001).

[13] Descartes frequently asserts that corporeal substance is divisible (AT 3, 477; CSMK, 202–203; AT 8A, 51, CSM 1, 231), even into parts that are really distinct from one another (AT 8A, 29, CSM 1, 213). But the fact that the different parts of matter are really distinct, and hence substances as such, does not imply that each part is an individual *body*. For Descartes, bodies are individuated by relative motion so that "if the division into parts occurs simply in our thoughts there is no resulting change" (AT 8A, 52; CSM 1, 232). Furthermore, and *pace* recent papers by Lennon (2007) and Schmaltz (2009) it is unclear why the real distinction among the parts of extension should threaten either the substantiality of extension itself or the doctrine that the parts of extension are mutually interdependent. On the first issue, there is no good Cartesian reason (I know of) to deny substantiality to something that has really distinct or substantial parts. On the second issue, the parts of extension seem to be mutually interdependent in the way I have described even granted all these parts are really distinct. The question is not whether the parts of extension of duration are really distinct – I think they are in both cases – but whether these parts are mutually interdependent in such a way that if one part exists then necessarily they all do.

are actually infinite" (AT 3 64; CSMK 147). Somewhat surprisingly, he extends this reasoning to our idea of indefinite space: "Similarly I could not conceive of an indefinite quantity by looking at a very small quantity or finite body unless the size of the world was actually or at least possibly indefinite" (Ibid.). My account offers a reason why Descartes is prepared to infer actually infinite space from the idea of finite extension: the parts of space are all so conceptually connected that the idea of any part requires the idea of all other parts, *ad indefinitum*.

The interdependence of the parts of space is also evident in Descartes's rejection of a plurality of worlds in *Principles* II, 22, which immediately follows his demonstration of indefinite extension. He says the matter composing any other world would have to be the same as composes this world, so not really a different world. This seems to leave him open to the pluralist objection that another world might be composed of the same kind of matter yet separate from this one. But he insists that this is impossible since "the matter whose nature consists simply in its being an extended substance already occupies absolutely all the imaginable space in which the alleged additional worlds would have to be located" (AT 8A; CSM 1 232). The suggestion seems to be that any 'other' extended world is conceived as bearing a spatial relation to the 'local' world, and therefore connected or interdependent in the way that the shorelines of two continents and the ocean between them all mutually connect.

So the parts of indefinite extension conceived as mere *res extensa* are mutually dependent because the parts are nothing more than their mutual geometrical relations. If one of two neighboring parts had failed to exist then another would have been in its place; but in that case the interloper would not be a distinct part from the original. And if *duration* were conceived purely abstractly or geometrically, perhaps as a line with a direction and flow, then its parts would be interdependent in a way similar to space. Reaching the 7th day of a week, for example, requires passing through the first 6 days. But, as we have seen, Descartes is at pains to distinguish the concrete duration of things (whether they move or not) from the abstract measure 'time'. The parts of duration, since they are identical to the continued existence of enduring things, are not conceptually interdependent. My brother might endure while I do not, and the world as a whole begins (or ends) when things begin (or cease) to endure. In response to Gassendi, Descartes makes clear that the independence of parts doctrine applies strictly to concrete duration rather than abstract time. Thus, when Gassendi complained against the Third Meditation proof: "Can we think of anything whose parts are more inseparable from one another than your duration?" Descartes responded that such interdependence characterizes at most "time considered in the abstract" but not "the duration of the thing which endures (*durantione rei durarantis*)" (AT 7 369-70; CSM 2 255). This explains why he ignores Gassendi's suggestion that the parts of a thing's duration "are merely external, they flow on without playing any active role" like a river flows past a rock (AT 7 301; CSM 2 209). For Descartes, such a metaphor wrongly takes duration (as opposed to time) to be ontologically separable from enduring things. But unlike the hours of a day conceived as a measure, or extension conceived purely geometrically, no part of my concrete duration is intrinsically linked to any other. And for this reason we cannot infer from what we imagine about the world beginning earlier that this is "truly imaginable, that is, real" (AT 8A 52; CSM 1 232).[14]

4.8 Temporal Infinitists

To summarize, Cartesian extension conceived purely abstractly or geometrically, independently of this or that fleeting body, must be indefinite (at least relative to our conception). In contrast, Cartesian duration is intrinsic to enduring minds and bodies and no part of this duration carries any guarantee of continuance or antecedence, however much we might imagine (or hope) it did. And this is the sense in which "natural reason" aids faith in proscribing a world without a temporal beginning or end (AT 11 656). It is worth noting that those among Descartes's contemporaries who privilege abstract time over concrete duration – that is, who think time is prior to and independent of the persistence of created things – tend to regard time as necessarily infinite; and they do so for reasons that parallel Descartes's reasons for denying infinity to concrete duration. For example, Gassendi insists that "time considered in itself has neither beginning nor end" (Brush 395). Gassendi is somewhat unclear why time in itself "cannot be stilled by any force", but he is fond of the analogy between space as mere extension in length, width and depth and time as a "flowing extension in which it is possible to designate past, present and future" (Brush 391). As we have seen, it is precisely this reification of time, conceived geometrically, that Descartes rejects.

Newton endorses Descartes's basic argument that "we cannot imagine a limit anywhere without imagining space beyond it" (DG 23). So he likewise agrees with the doctrine of *Le Monde* that the concept of indefinite space is presupposed in the idea of any finite object: "if any being whatsoever is posited, space is posited" (DG 25). Unlike Descartes, Newton extends this analysis to duration: "we cannot think that there is no duration even though we might suppose that no thing endures" (DG 26). Like Descartes's abstract time, the parts of Newton's duration depend on their relations to all other parts: "the parts of duration are individuated by their order; if for example yesterday could change places with today, it would lose its identity and become today" (DG 25). We can, of course, easily imagine an end to the duration of this or that enduring thing, as Descartes points out. But it is as hard to conceive an end to a mere ordering apart from anything else – such as the *Principia*'s "absolute, true, and mathematical time, [which] of itself, and from its own nature, flows equa-

[14] I do not here take up the question of the infinite or indefinite divisibility of time. In my view, the same considerations in favor of the indefinite division of space carry over to duration. This is not inconsistent with the independence of parts doctrine since that concerns the relation between successive parts of duration not the proper sub-parts of any temporal interval. The discussion of Cartesian temporal (dis)continuity has a long history. Recent examples: Arthur (1988), Garber (1992), Levy (2005), and Gorham (2008b).

bly without relation to anything external" (Principia 408) – as to conceive a limit on geometrical space or the series of natural numbers.

Locke holds that anyone "may easily conceive in his mind the beginning of motion but not at all duration. . . so also he may set limits to body, and the extension belonging to it, but not to space" (Essay II, xiv, 26; 193). But, even more than Gassendi and Newton, Locke recognizes the role mathematical abstraction, particularly number, plays in convincing us of the necessary infinity of space and time: "in space and time, when the mind pursues the idea of infinity it there makes use of the ideas and repetitions of number" (Essay II, xvii, 9; 215). More specifically, we think of the infinite series of natural numbers as "like a line, whereas one end terminating with us, the other is extended still forward beyond all that we can conceive" and "in duration, we consider it as if this line of numbers were extended both ways to an unconceivable, indeterminate and infinite length" (Essay II xvii 10; 215).

Gassendi, Newton and Locke all closely associate infinite space and time with God's immensity and eternity. But Descartes strongly rejects any literal conception of God's immensity: "the alleged extension of God cannot be the subject of the true properties which we perceive very distinctly in all space" (AT 5 271; CSMK 362).[15] But shouldn't the eternal duration of God (along with the eternal truths he decreed) entail that time did not begin? Descartes himself suggests as much to Burman in explaining God's eternity:

> since we can divide it up after the creation of the world why should it not have been possible to do the same before creation since duration remains constant? Thus, eternity has now co-existed with created things for, say, five thousand years, and occupied time; so it could have done the same before creation, if we had had some standard to measure it by. (AT 5 149; CB 7)

However, I believe the passage can be reconciled with the letter to Chanut quite easily. Since God is eternal (i.e. sempiternal), duration is indeed beginningless. But this follows from the nature of God, not from our conceptions of time, whereas the indefinite extension of the universe follows simply from our conception of space. It is perhaps less easy to reconcile the Burman passage with Descartes's dismissal as contradictory More's speculations about a 'between-world' duration (CSMK 373).[16] But recall that Descartes is here responding specifically to More's suggestion that the between-world (*intermundium*) or 'world-absence' (*absentia mundi*) "would have its own duration" (AT 5 302). The contradiction consists in supposing that the 'world-absence' itself would have duration, any more than it would have extension, not in supposing that God would then endure. The point to Chanut remains the same: the universe is by its nature indefinitely extended but its duration depends on the arbitrary will of God.

---

[15] See also AT 5 275; CSMK 364; AT 5 342; CSMK 372.

[16] Koyré asserts that Descartes rejects between-world duration because it would mean "introducing time into God" (1957, 122). But why should duration between worlds make God temporal, any more than during worlds? Furthermore, there is no reason to regard a temporal God as 'contradictory' for Descartes and very good reasons to suppose that his God was in fact temporal. See Gorham (2008a).

4.9 Descartes and Early Spinoza

In early works of Spinoza, the influence of Descartes's metaphysics of time is very clear. In the 'metaphysical thoughts' (*Cogitata Metaphysica*), which he appended to his 1663 'synthetic' exposition of Descartes's *Principles of Philosophy* (*Descartes Principiorum Philosophiae*), Spinoza defines duration as "the attribute under which we conceive the existence of created things insofar as they persevere in their actuality" (S 104; G 1, 244). So conceived, he notes, duration is distinct only in reason from the total existence of a thing since "as much as you take away from the duration of thing so much necessarily you take away from its existence" (S 104-5; G 1 244). As for time, "in order that duration may be determined, we compare it with other things that have a fixed and determinate motion, and this comparison is called time" (S 105; G I 244). Such 'clock time', he emphasizes, "is not an affection of things. . . but rather a mode of thinking (*modus cogitatndi*) that we use to explicate duration" (Ibid). In all of this he follows Descartes closely: the distinction between duration and time; the identification of the former as an attribute (continuation in existence) that is merely distinct from the enduring thing; and that classification of the latter as a conventional measure and hence a mere mode of thought. And so Spinoza likewise agrees with Descartes that duration (and time) ceases "when created things cease to exist and begin when created things begin to exist" (S 129; G I 169).

But in a contemporaneous, independent letter ("On the Infinite"), Spinoza retains the distinction between time and duration (and eternity) while seeming to hold that duration is infinite. He associates duration with the existence of modes, which he suggests are infinite by the "force of the cause in which they inhere" (G IV 61, 1-7; C 205). By this he seems to mean that the duration of modes is infinite insofar as "they flow from eternity without which they cannot be rightly understood" (G IV 58, 2-3; C 203). What is noteworthy for our purposes is the duration of modes is conceived as finite precisely because we abstractly divide it into parts: "when they are conceived abstractly they can be divided into parts and regarded as finite" (G IV 61 2-3; C 205). Such division is the source of Zeno-like paradoxes of infinity: "when someone has conceived Duration abstractly, and by confusing it with time begun to divide it into parts, he will never be able to understand how an hour can pass" (G IV 58 5-7; C 203).

So the early Spinoza seems to split from Descartes on the question of infinite duration precisely because he does not attribute parts to duration. This disagreement is evident in his generally sympathetic exposition of Descartes's *Principles of Philosophy*. For instance, in his exposition of the second causal proof of God he conspicuously omits the crucial premise about the mutual independence of the parts of time. Spinoza's critical exposition of the proof makes it instead depend entirely on the assumption that if I were self-created, I would have given myself every perfection. Similarly, Descartes's proof of the law of rectilinear motion turns crucially, though somewhat obscurely, on the assumption that God preserves motion "in the precise form in which it is occurring at the very moment he preserves it" (CSM I 242; AT 8A 64). But Spinoza's puzzling reconstruction proceeds entirely from the 'nature of motion' which supposedly excludes from consideration any "duration that can be conceived as greater than another duration" (G I, 204; S 65). These amendments to Descartes's official proofs illustrate Spinoza's opposition to Descartes on a metaphysical principle – the reality of temporal parts – that had major implications for their respective systems.

4.10 Conclusion

Descartes is a crucial transitional figure in the revolutionary transformation of space/time concepts in the seventeenth century. Through his doctrine of *res extensa*, he reified the indefinite imaginary space of the late scholastics: the space we imagine beyond the world is no different from the matter we perceive in the world itself. Although the space of the early *Le Monde* is a plenum rather than a vacuum, it anticipates many of the crucial attributes of Newton's absolute space: boundless, purely geometrical, continuous, independent of any other being, and clearly conceived. Descartes's view of time was also proto-absolutist insofar as he divorced successive duration from motion and its measure. He was not, however, ready to reify imaginary time. A time independent of everything else besides God, before and after creation, for example, would qualify as a genuine substance. But such a substance would fall outside the dualism of mind and body: neither thinking nor extended in three dimensions. If we stick to the actual world of minds and bodies, duration is best regarded as a universal (transcendental) attribute; if we abstract time from existing things, it is a mere mode of thought. His asymmetrical attitude about indefinite space vs. time therefore seems, like so many Cartesian doctrines, to be a consequence of his strict dualism.

Abbreviations

AT	Descartes, R. 1983. *Oeuvres De Descartes*, 11 vols. Ed. C. Adam and P. Tannery. Paris: J. Vrin. Citation by volume and page.
Brush	Gassendi, P. 1972. *Selected Works of Pierre Gassendi*. Ed. C. Brush. Johnson Reprint Co. Citation by page
C	Spinoza, B. 1985. *The Collected Works of Spinoza*, Vol. I. Ed. E. Curley. Princeton: Princeton University Press. Citation by page.
CB	Descartes. R. 1976. *Descartes' Conversation with Burman*. Ed. J. Cottingham. Oxford: Clarendon Press. Citation by page.
CSM	Descartes, R. 1984–5. *The Philosophical Writings Of Descartes*, 2 vols., Trans. J. Cottingham, R. Stoothoff, and D. Murdoch. Cambridge: Cambridge University Press. Citation by volume and page number.
CSMK	Descartes, R. 1991. *The Philosophical Writings Of Descartes: The Correspondence*. Trans. J. Cottingham, R. Stoothoff, D. Murdoch, and

	A. Kenny. Cambridge: Cambridge University Press. Citation by page number.
DM	Suárez, F. 1866. *Disputationes Metaphysicae*. In *Opera Omnia*. Ed. by C. Berton. Paris: Vives. Citation by disputation, section and paragraph.
DG	Newton, I. 2004. *De Gravitatione et aequipondio fluidorum.* In *Newton: Philosophical Writings*. Ed. and Trans. A. Janiak and C. Johnson. Cambridge: Cambridge University Press. Citation by page number.
Essay	Locke, J. 1975. *An Essay Concerning Human Understanding*. Ed. P. H. Nidditch. Oxford: Oxford University Press. Citation by book, chapter, section and page number.
G	Spinoza. B. 1925 *Opera,* IV Vols. Ed. C. Gebhardt. Heidelberg: Carl Winter. Citation by volume and page number.
Principia	Newton, I. 1999. *The Principia: Mathematical Principles of Natural Philosophy*. Ed. and Trans. I.B. Cohen and A. Whitman. Berkeley. Citation by page number.
S	Spinoza, B. 1998. *The Principles of Cartesian Philosophy* and *Metaphysical Thoughts*. Trans. S. Shirley. Indianapolis: Hackett Publishing Co. Citation by page number.

Acknowledgements For helpful comments, I would like to thank Edward Slowik and the participants in the New York Seminar in Early Modern Philosophy, Jerusalem, June 2016.

References

Ariew, R. (1987). The infinite in Descartes' conversation with Burman. *Archiv für Geschichte der Philosophie, 69*, 140–183.
Arthur, R. (1988). Continuous creation, continuous time: A refutation of the alleged discontinuity of Cartesian Time. *Journal of the History of Philosophy, 26*, 349–375.
Bexley, E. (2012). Quasi-absolute time in Francisco Suárez's metaphysical disputations. *Journal of the History of Ideas, 22*, 5–22.
Daniel, S. (1981). Seventeenth-century scholastic treatments of time. *Journal of the History of Ideas, 42*, 587–606.
Des Chene, D. (1996). *Physiologia*. Ithaca: Cornell University Press.
Duhem, P. (1987). *Medieval cosmology: Theories of infinity, place, time, void and the plurality of worlds*. Chicago: University of Chicago Press.
Garber, D. (1992). *Descartes' metaphysical physics*. Chicago: University of Chicago Press.
Gorham, G. (2004). Cartesian causation: Continuous, instantaneous, overdetermined. *Journal of the History of Philosophy, 42*, 389–423.
Gorham, G. (2007). Descartes on time and duration. *Early Science and Medicine, 12*, 28–54.
Gorham, G. (2008a). Descartes on God's relation to time. *Religious Studies, 44*, 412–431.
Gorham, G. (2008b). Cartesian temporal atomism: A new defence, a new refutation. *British Journal for the History of Philosophy, 16*, 625–637.
Gorham, G. (2012). 'The twin-brother of space': Spatial analogy in the emergence of absolute time. *Intellectual History Review, 22*, 23–39.
Grant, E. (1981). *Much Ado about nothing*. Cambridge: Cambridge University Press.
Kendrick, N. (1998). 'Uniqueness in Descartes' 'Infinite' and 'Indefinite'. *History of Philosophy Quarterly, 15*, 23–36.

Koyré, A. (1957). *From the closed world to the infinite universe*. Baltimore: Johns Hopkins University Press.

Lennon, T. (2007). The eleatic Descartes. *Journal of the History of Philosophy, 45*, 29–47.

Levy, K. (2005). Is Descartes a temporal atomist? *British Journal for the History of Philosophy, 13*, 627–674.

LoLordo, A. (1997). *Pierre Gassendi and the birth of early modern philosophy*. Cambridge: Cambridge University Press.

Schmaltz, T. (2009). Descartes on the extensions of space and time. *Analytica, 13*, 113–147.

Slowik, E. (2001). Descartes and individual corporeal substance. *British Journal for the History of Philosophy, 9*, 1–15.

Wilson, M. (1999). Can I be the cause of my ideas of the world? (Descartes on the infinite and indefinite). In A. O. Rorty (Ed.), *Ideas and mechanism: Essays on early modern philosophy* (pp. 339–358). Princeton: Princeton University Press.

Chapter 5
"A Substance Consisting of an Infinity of Attributes": Spinoza on the Infinity of Attributes

Yitzhak Y. Melamed

Abstract At the opening of the *Ethics* Spinoza defines God as a substance consisting of infinitely many attributes. Still, the reader of the *Ethics* will find only two of these attributes discussed in any detail in Parts Two through Five of the book. Addressing this intriguing gap between the infinity of attributes asserted in E1d6 and the discussion of merely the two attributes of Extension and Thought in the rest of the book, Jonathan Bennett writes: "Spinoza seems to imply that there are other [attributes] – he says indeed that God or Nature has "infinite attributes." Surprising as it may seem, there are reasons to think that by this *Spinoza did not mean anything entailing that there are more than two attributes*." In this paper I show that Bennett's claim is fundamentally wrong and deeply misleading. I do think, however, that addressing Bennett's challenge helps us better understand Spinoza's notion of infinity. I begin by summarizing Bennett's arguments and then turn to examine briefly the textual evidence for and against his reading. I respond to each of Bennett's arguments, and conclude by pointing out some theoretical considerations that, I believe, simply refute his reading.

5.1 Introduction

At the beginning of Part I of the *Ethics*, Spinoza provides the following definition of God:

> E1d6: By God I understand a being absolutely infinite, i.e., a substance consisting of an infinity of attributes, of which each one expresses an eternal and infinite essence [*Per Deum intelligo ens absolute infinitum, hoc est, substantiam constantem infinitis attributis, quorum unumquodque aeternam, et infinitam essentiam exprimit*].

> Exp.: I say absolutely infinite, not infinite in its own kind; for if something is only infinite in its own kind, we can deny infinite attributes of it [NS (i.e., we can conceive infinite

Y. Y. Melamed (✉)
Philosophy Department, Johns Hopkins University, Baltimore, MD, USA
e-mail: ymelame1@jhu.edu

O. Nachtomy, R. Winegar (eds.), *Infinity in Early Modern Philosophy*, The New Synthese Historical Library 76,
https://doi.org/10.1007/978-3-319-94556-9_5

> attributes which do not pertain to its nature)]; but if something is absolutely infinite, whatever expresses essence and involves no negation pertains to its essence [*Explicatio. Dico absolute infinitum, non autem in suo genere; quicquid enim in suo genere tantum infinitum est, infinita de eo attributa negare possumus; quod autem absolute infinitum est, ad ejus essentiam pertinet, quicquid essentiam exprimit, et negationem nullam involvit*].[1]

The definition of God and its *explicatio* draw an important distinction between what is *absolutely infinite* and what is merely *infinite in its own kind*. God is absolutely infinite, but each of the infinitely many attributes of God is merely infinite in its own kind,[2] since we can deny of it all of the *other* infinitely many attributes. Spinoza makes the last point quite explicitly in one of his early letters, where he uses the example of the attribute of extension and states: "Extension is not infinite absolutely, but only insofar as it is Extension, i.e., in its own kind."[3]

Though this most fundamental definition of the *Ethics* unequivocally asserts that God has infinitely many attributes, the reader of the *Ethics* will find only two of these attributes discussed in any detail in Parts Two through Five of the book. Addressing this intriguing gap between the infinity of attributes asserted in E1d6 and the discussion merely of the two attributes of Extension and Thought in the rest of the book, Jonathan Bennett writes:

> Spinoza seems to imply that there are other [attributes] – he says indeed that God or Nature has "infinite attributes." Surprising as it may seem, there are reasons to think that by this *Spinoza did not mean anything entailing that there are more than two attributes*.[4]

In this paper I will argue that Bennett's claim is fundamentally wrong and deeply misleading. I do think, however, that addressing Bennett's challenge will help us better understand Spinoza's notion of infinity. I will begin by summarizing Bennett's arguments. I will then turn to examine briefly the textual evidence for and against his reading. Then I will respond to each of Bennett's arguments, and conclude by pointing out theoretical considerations which, I believe, simply refute his reading.

---

[1] **NS** [*Nagelate Schriften*] refers to the text of the 1677 Dutch translation of Spinoza's works. Unless otherwise marked, all quotes from Spinoza's works and letters are from Curley's translation. I have relied on Gebhardt's critical edition (*Spinoza Opera*, 4 volumes (Heidelberg: Carl Winter Verlag, 1925)) for the Latin text of Spinoza. I use the following standard abbreviations for Spinoza's works: **TdIE** - *Treatise on the Emendation of the Intellect* [*Tractatus de Intellectus Emendatione*], **DPP** – *Descartes's Principles of Philosophy* [*Renati des Cartes Principiorum Philosophiae Pars I & II*], **CM** – *Metaphysical Thoughts* [*Cogitata Metaphysica*], **KV** – *Short Treatise on God, Man, and his Well-Being* [*Korte Verhandeling van God de Mensch en deszelfs Welstand*], **TTP** –*Theological-Political Treatise* [*Tractatus Theologico-Politicus*], **Ep.** – *Letters*. Passages in the *Ethics* will be referred to by means of the following abbreviations: a(-xiom), c(-orollary), p(-roposition), s(-cholium) and app(-endix); 'd' stands for either 'definition' (when it appears immediately to the right of the part of the book), or 'demonstration' (in all other cases). Hence, E1d3 is the third definition of part 1 and E1p16d is the demonstration of proposition 16 of part 1.

[2] See Spinoza's assertion in E1p16d: "each of the attributes expresses an essence infinite *in its own kind*" (italics added).

[3] Ep. 4| IV/13/2-21.

[4] Bennett, "Spinoza's Metaphysics," 65. Italics added.

5.2 Bennett on the Infinity of the Attributes

Bennett presents the following five arguments to motivate his surprising claim: (1) Spinoza frequently uses "infinite" as virtually synonymous with "all." The claim that God has all the attributes merely commits him to the view that whatever attributes are there, they must be instantiated in God. If there are only two possible attributes, then the claim that God has infinite attributes amounts to nothing over and above the claim that God has two attributes.[5] (2) If Spinoza was serious in ascribing infinitely many attributes to God, he should have discussed them in some detail in the body of the *Ethics*.[6] (3) In Letters 64 and 66, Spinoza attempts to explain why we cannot know any attributes other than thought and extension. However, argues Bennet, Spinoza's claim is "a move so abrupt, *ad hoc*, and unexplained that we cannot even be sure whether it is a retraction of the metaphysics or of the epistemology."[7] (4) The traditional conception of God as an *ens realissimum* could have motivated Spinoza to ascribe to God *all* attributes or perfections. However, there was no respectable theological tradition that would motivate him to ascribe to God *infinitely* many attributes.[8] (5) Spinoza had no theoretical or philosophical pressure that would push him to assert that God has more than two attributes.[9]

5.3 The Textual Evidence

Let us turn now to examining Spinoza's text and check whether it can support the claim that God/Nature has no more than the two attributes of extension and thought. We'll begin with a simple question: is there *any* text in Spinoza's *oeuvre* in which Spinoza asserts that there are no more than two attributes? To the best of my knowledge, the answer is "no."[10] In contrast, we have an abundance of texts – in the *Ethics* and outside it – in which Spinoza clearly commits himself to the existence of

---

[5] Bennett, *Study*, 76; Bennett, "Spinoza's Metaphysics," 65.

[6] Bennett, *Study*, 78–79.

[7] Bennett, *Study*, 78.

[8] Bennett, *Study*, 76–77; Bennett, "Spinoza's Metaphysics," 66.

[9] Bennett, Study, 77.

[10] The closest Spinoza comes to the last claim is in a note to the first chapter of the first part of the *Short Treatise* where he claims: "After the preceding reflections on Nature we have not yet been able to find in it more than two attributes that belong to this all-perfect being." (I/17/35-38). However, he *immediately* continues to argue against the view that God has merely two attributes: "And these give us nothing by which we can satisfy ourselves that these would be the only ones of which this perfect being would consist. On the contrary, we find in ourselves something which openly indicates to us not only that there are more, but also that there are infinite perfect attributes which must pertain to this perfect being before it can be called perfect." (I/17/38-42).

attributes other than thought and extension. Consider the following two passages from E2p7s:

> Whether we conceive nature under the attribute of Extension, or under the attribute of Thought, *or under any other attribute*, we shall find one and the same order, *or* one and the same connection of causes, i.e., that the same things follow one another.[11]

> So long as things are considered as modes of thinking, we must explain the order of the whole of nature, or the connection of causes, through the attribute of Thought alone. And insofar as they are considered as modes of Extension, the order of the whole of nature must be explained through the attribute of Extension alone. *I understand the same concerning the other attributes*.[12]

Notice that in both passages Spinoza does not entertain *the slightest doubt* about the existence of the unknown attributes. He does not say, "I understand the same concerning the other attributes, *if there are any*." Instead, he affirms *without any reservation* that the same order of explanation should obtain with regard to the other, unknown attributes. Interestingly, we do have a nice example of Spinoza's formulating a claim about an issue he is not confident about. Consider E3p2: "The Body cannot determine the Mind to thinking, and the Mind cannot determine the Body to motion, to rest or to anything else (*if there is anything else*)."[13] In E3p2 Spinoza reservedly entertains the possibility that a body *might* be determined to states other than motion and rest, without committing himself to the existence of this third kind of state. In contrast, both passages in E2p7s clearly commit Spinoza to the existence of attributes other than extension and thought.

In addition to the two crystal-clear passages from E2p7s, there is an interesting yet more intricate passage in E2p13d, in which Spinoza is bothered by the possibility of a mismatch between the minds of modes of different attributes, i.e., he is bothered by the possibility that "the object of the Mind were something else also, in addition to the Body."[14] In order to rule out the possibility that my mind might have as its object not only my body but also a mode of one of the unknown attributes, Spinoza appeals to E2a5, which asserts that the human mind has access *only* to modes of extension and thought. I reconstruct this argument in greater detail in another place.[15]

Turning now to Spinoza's correspondence, in Letter 56 (dated Oct./Nov. 1674) Spinoza writes:

> I don't say that I know God completely, but only that I know some of his attributes, not all of them, *not even most of them. Certainly, being ignorant of most of them*, does not prevent my knowing some.[16]

---

[11] II/90/14-18. Italics added.

[12] II/90/23-28. Italics added.

[13] Italics added.

[14] II/96/12.

[15] Melamed, *Spinoza's Metaphysics*, 169–171.

[16] IV/261/11-15. Italics added.

Spinoza's assertion that we do not know the *majority* of the attributes clearly entails that he believed there are at least... five attributes. In an earlier letter, Spinoza referred to "other attributes" of God other than intellect (i.e. thought),[17] hence implying that there must be at least... three attributes.[18]

In Letters 64 and 66 Spinoza unmistakably asserts the existence of infinitely many attributes unknown to the human mind. We will postpone the explication of these crucial letters until we address the important question of Spinoza's reasons for asserting that we cannot know any attributes other than thought and extension.

In the *Theological Political Treatise*, Spinoza does not employ his typical metaphysical terminology of substance, attributes, and modes[19]; yet in a note appended to his discussion of nature in the sixth chapter, he remarks: "By Nature here I understand not only matter and its affections, but in addition to matter, infinite other things [*alia infinita*]."[20] It is highly likely that these "infinite other things" are the infinite attributes (apart from extension).

In the *Short Treatise*, the infinitely many unknown attributes are discussed in detail in KV I/1,[21] and the Second Appendix to the *Short Treatise* offers an elaborate discussion of the nature of the minds, or souls, of the modes of the unknown attributes:

> The essence of the soul consists only in the being of an Idea, or objective essence, in the thinking attribute, arising from the essence of an object which in fact exists in Nature. I say *of an object that really exists,* etc., without further particulars, in order to include here not only the modes of extension, but also the modes of all the infinite attributes, which have a soul just as much as those of extension do.[22]

Finally, as a piece of external evidence, consider the following passage from Leibniz's notes on Spinoza's metaphysics, following a conversation he had with their mutual friend, Walter von Tschirnhaus. As we will shortly see, Tschirnhaus was particularly knowledgeable about Spinoza's views on the unknown attributes:

> *He [Spinoza] thinks that there are infinite other positive attributes besides thought and extension.* But in all of them there is thought, as here there is in extension. What they are like is not conceivable by us; every one is infinite in its own kind, like space here.[23]

In summary, we have, I believe, a solid body of textual evidence committing Spinoza to the existence of infinitely many other attributes beyond thought and

---

[17] Ep. 35| IV/181/16.

[18] Though I would not put much weight on the last source, since it addresses Spinoza's 1663 book, *Descartes' Principles of Philosophy*, and one could argue that there Spinoza is referring to the Cartesian conception of God's attributes.

[19] See Melamed, "Glimpse into Spinoza's Metaphysical Laboratory," 277–278.

[20] TTP Ch. 6| III/83/10,

[21] See note d| I/17/34-48.

[22] KV Appendix II| I/119/6–13. Cf. I/120/1-6.

[23] Pollock, *Spinoza*, 161. Leibniz, *Sämtliche Schriften*, 6th Series, Volume 3, 385 (lines 12–15). Italics added. I slightly amended Pollock's translation by replacing 'in this world' by 'here', which is more loyal to the Latin *hic*.

extension. We do not have even a single text in which Spinoza asserts that God has, or even might have, only the two known attributes of extension and thought.

5.4 Responses to Bennett's Arguments

I turn to address Bennett's arguments in favor of his reading. (1) Bennett's claim that in Spinoza 'infinite' always means 'all' is not precise, to say the least. In Letter 12, the foremost text for Spinoza's discussion of infinity, he notes that some "kind of infinite can be conceived to be greater than another infinite, without any contradiction."[24] Thus, the smaller infinity may not contain *all* the items contained within the larger infinity. It is also clear that in the explication of the definition of God in the *Ethics*, 'infinite' is *not* synonymous with 'all.' At the beginning of this paper, we have seen that each attribute is "infinite in its own kind." Thus, when Spinoza claims in E1d6e "if something is only infinite in its own kind, we can deny infinite attributes of it," he cannot mean that we can deny *all* attributes of what is infinite in its own kind; if we deny *all attributes* from, say, the attribute of extension (i.e., what is infinite in its own kind), nothing would be left.

Thus, it seems that in the two most significant texts in which Spinoza discusses infinity, he uses 'infinity' not as synonymous with 'all'. But this is a relatively minor point. The following consideration seems to me more important. Bennett is right in claiming that if by 'infinity' Spinoza meant nothing over and above 'all', and if there were only two possible attributes, then for God to have infinite attributes would amount to nothing more than having two attributes. Yet, why not extend Bennett's logic one step further? If *no* attributes were possible (and *no* modes were possible), it would still be perfectly correct under Bennett's reading to assert that God has infinitely many attributes and infinitely many modes. If no attributes are possible, then God may still have all attributes, even though he has none (i.e., it would still be trivially true that he has all possible attributes). But since – per Bennett's reading – 'infinite' is synonymous with 'all', one could then also truly assert that God has infinitely many attributes (though he has none). Yet, such an assertion would be highly misleading, and it would make no sense for a speaker who even suspects that there might be no possible attributes to assert that there are infinitely many attributes. The very same consideration also works against reading Spinoza's infinity of attributes as "all, that is, two." If Spinoza meant to claim that God merely has all the attributes, why should he not use the simple and common term 'all' instead of the highly misleading 'infinity'?

(2) Spinoza does not elaborate upon the nature of the other attributes in Parts Two to Five of the *Ethics* for a simple reason: the aim of these parts is the study of the nature of the human mind and the best measures leading to its blessedness. The preface to Part Two announces as much. It explains that from that point onward,

[24] Ep. 12| IV/53/12-14.

Spinoza is homing in on a tiny fraction of his universe – the fraction that is relevant to the knowledge and blessedness of the human mind:

> I pass now to explaining those things which must necessarily follow from the essence of God, or the infinite and eternal Being—not, indeed, all of them, for we have demonstrated (IP16) that infinitely many things must follow from it in infinitely many modes, *but only those that can lead us, by the hand, as it were, to the knowledge of the human Mind and its highest blessedness.*[25]

From Part Two onward Spinoza is focusing on the restricted part of his universe that is relevant to the achievement of human blessedness. For the most part, the knowledge of the infinitely many unknown attributes is irrelevant to this endeavor.[26] This is the trivial and primary reason for Spinoza's silence about the infinitely many attributes.[27] In addition, Spinoza had good reasons to believe that while we know that God/Nature has infinitely many attributes, we can hardly know anything about the nature of these attributes. He develops these claims in Letters 64 and 66, and we shall turn now to examine his reasons.

(3) Spinoza had a perfect explanation for the fact that one does not know the nature of any attributes other than thought and extension. According to Spinoza, the human mind is a complex idea (i.e., mode of Thought) whose object is nothing but a human body (a mode of Extension).[28] One of the most central doctrines of the *Ethics* asserts that there is a parallelism, or isomorphism, between the order of things and the order of ideas (E2p7).[29] Things (*res*) for Spinoza are everything that is real, including bodies and ideas. In E1p10, Spinoza argues that insofar as an attribute constitutes the essence of substance (per E1d4), each attribute must be conceived through itself, just like the substance. E1p10 thus establishes a *conceptual barrier* between the attributes. Relying on E1p10 (and E1a4), Spinoza proves in E2p6 that the attributes are also *causally* isolated from each other (i.e., a mode from one attribute cannot cause a mode from another attribute). Thus, there is a causal and conceptual barrier between the infinitely many attributes.[30]

In Ep. 66, Spinoza relies on these two doctrines – the Ideas-Things Parallelism of E2p7[31] and the barrier among the attributes of E1p10 – to prove that not only can items belonging to different attributes not interact causally with each other, but also *mental representations of items belonging to different attributes cannot causally*

[25] E2prefl II/84/8-12. Italics added.

[26] Though, as we have already seen, the issue occasionally crops up in E2p13s, given the possibility of a mismatch between minds and their proper object.

[27] One may speculate that the version of the *Ethics* written by Spinoza's twin in the third attribute would be silent about the nature of extension, since the latter kind of knowledge would be of no use in studying the measures leading to the blessedness of the mind of this third-attribute twin of Spinoza.

[28] E2p13.

[29] E2p7: "The order and connection of ideas is the same as the order and connection of things."

[30] See Della Rocca's elegant account of the barrier in *Representation*, 9–22.

[31] For a detailed explication of the Ideas-Things Parallelism, see Melamed, *Spinoza's Metaphysics*, Chapter Five.

interact with each other. In other words, in addition to the barrier among the attributes introduced in E1p10 and E2p6, there is a *parallel barrier within* the attribute of thought among representations (i.e. ideas) whose objects are items belonging to different attributes. Thus, it is not only the case that my body cannot causally interact with a mode of the third attribute, but also the case that my mind (which is just the idea of my body) cannot causally interact with any mind (or idea), which represents items of the third attribute. The *parallel barrier*, which is internal to Thought, does not allow *any communication* between ideas representing different attributes. Our minds (i.e., the ideas of our bodies) cannot communicate with the minds of the (modes of the) infinitely many unknown attributes, just as our bodies cannot interact with the modes of the infinitely many other attributes. Each attribute (and its representation in thought) is isolated from every other attribute (and its representation in thought). Thus, contrary to Bennett's claim, Spinoza's argument in Letters 64 and 66 is well-grounded in E1p10 and E2p7.[32]

(4) Spinoza was well acquainted with a philosophical and theological tradition that ascribes infinitely many attributes to God, though it was not the tradition under Bennett's spotlight. In his discussion of the divine attributes in the *Light of the Lord*, Crescas develops in great detail the claim that God has infinitely many attributes and that each of his attributes is infinite.[33] Given Spinoza's detailed discussion and endorsement of Crescas' conception of actual infinity in Letter 12,[34] it is highly unlikely that he was unaware of this claim, especially since Crescas was not the only medieval Jewish thinker to advance such an argument.[35] Another philosopher with whom Spinoza was of course acquainted and who affirmed that God has "countless" attributes beyond the ones we know is Descartes.[36] These "countless" attributes of God cannot be just non-essential modes, since Descartes explicitly denies that God has any modes.[37] There is, however, a subtle yet important difference between these claims of Descartes and Spinoza. Descartes's claim that there are uncountable divine attributes which we cannot comprehend secures the transcendence of the Cartesian God.[38] Spinoza's claim that *Deus sive Natura* has infinitely many attributes which are not accessible to us makes *Nature* (with capital N, i.e., as not restricted to extended and thinking nature) *just as transcendent to us as God is*.[39]

---

[32] For a detailed presentation of this issue, see Melamed, *Spinoza's Metaphysics*, Chapter Six, and Melamed, "Spinoza's Metaphysics of Thought."

[33] Crescas, *Or ha-Shem*, I, 3, 3 (pp. 106–108 in the 1990 edition). Cf. Harvey, *Rabbi Hasdai Crescas*, 91–94, and Levy, *Figures de l'infini*, 204–207.

[34] Ep. 12| IV/62/1-10.

[35] See Harvey, *Rabbi Hasdai Crescas*, 94.

[36] See Descartes's Letter to Mersenne from July 1641 (AT III 394| CSM III 185).

[37] See *Principles of Philosophy*, I 56. Cf. *Comments on a Certain Broadsheet* (AT VIIIB 348| CSM I 297).

[38] On the incomprehensibility of the infinite in Descartes, see Ariew, "The Infinite," 17.

[39] Notice, however, that in contrast to the incomprehensibility of the Cartesian God, Spinoza's "unknown attributes" are comprehended by some *finite* minds, i.e., the minds of the modes of the unknown attributes.

This is a bold and highly original view which is consistent with Spinoza's deep critique of anthropocentrism.[40]

(5) Let's turn to Bennett's final point. Were there any theoretical and philosophical pressures within Spinoza's system that would push him to affirm the existence of more than two attributes? Yes, there were. We will point out two strong reasons that motivated Spinoza to affirm that God has infinitely many attributes beyond extension and thought.

(i) Spinoza's first reason for affirming that God has infinitely many attributes beyond thought and extension is quite straightforward. Both in the *Short Treatise* and in the *Ethics* Spinoza denies that the infinite can be composed of an accumulation of finite parts.[41] Now let's suppose, per Bennett's suggestion, that there is a finite number, *n*, of possible attributes. Thus, "the infinity of attributes," i.e., the number *n* of attributes, would be composed of *n* attribute units, and this would flatly contradict Spinoza's assertion that the infinite cannot be composed from the finite. Put in other words, Bennett's suggestion that for Spinoza 'infinite' is synonymous with 'all' makes Spinoza's notion of infinity nothing but the sum-total of possible attributes. However, Spinoza explicitly rejects the view that infinity is the sum of any aggregation of finite units.
(ii) To begin elucidating Spinoza's second reason, consider E1p9:

> The more reality or being [*esse*] each thing has, the more attributes belong to it.

In E1p10s, Spinoza points to E1p9 as explaining his reason for defining God – at the very beginning of Part One – as consisting of an infinity of attributes.

> Nothing in nature is clearer than that each being must be conceived under some attribute, and the more reality, or being it has, the more it has attributes which express necessity, *or* eternity, and infinity. And consequently, there is also nothing clearer than that a being absolutely infinite must be defined (as we taught in D6) as a being that consists of infinite attributes, each of which expresses a certain eternal and infinite essence.[42]

The passage above would appear pretty odd under Bennett's reading: why would Spinoza formulate a *general rule* about the correspondence between the reality and the number of attributes a thing has, when only two attributes are possible at all? Still, to expose oddity is not to refute. Fortunately, we can push this line of objection toward a more conclusive result.

In a letter dated October 1674, 3 years before his death, Spinoza writes:

> Truly, I confess I still don't know in what respect spirits are more like God than other creatures are. *I know this: that there is no proportion* [*nullam esse proportionem*] *between the*

[40] See my paper, "Spinoza's Anti-Humanism."

[41] "The infinite cannot be composed of a number of finite parts" (KV I 1 |I/18/10). E1p15s (II/58/27): "infinite quantity... is not composed of finite parts." Cf. KV II 24 |I/107/1: "For how is it possible that we could infer an infinite and unlimited thing from one that is limited?"

[42] E1p10s| II/72/10-17. Italics added.

> *finite and the infinite*; so the difference between the greatest, most excellent creature and God is the same as that between the least creature and God.[43]

Since the reality of God (per E1p9) correlates with the number of attributes God has, then, if God were to have any finite number of attributes *n*, there would be a clear and simple proportion between the reality of a finite being – for example, me – and God's reality. Since I am constituted by modes of *two* attributes, the proportion between God's reality and mine would be *precisely*: *n*/2. Yet, as the passage above states unequivocally, Spinoza denies the very possibility of such a ratio between the infinity of God and the finitude of finite things.[44]

5.5 Conclusion

Toward the very end of his celebrated 1663 "Letter on the Infinite," Spinoza writes:

> I have, finally, set out briefly the causes of the errors and confusions which have arisen concerning the Problem of the Infinite, and unless I am mistaken, I have so explained all of them that I do not think any Problem about the Infinite remains which I have not touched on here or which cannot be solved very easily from what I have said. So I don't regard it as worthwhile to detain you any longer with these matters.[45]

The "Letter on the Infinite" is a difficult yet extremely important text for the study of Spinoza's metaphysics, as it contains a concise presentation of the core of his metaphysics, as well as Spinoza's advocacy of actual infinity.[46] Still, Spinoza's boasting – in the passage above – that he provided a complete explanation of all the aspects of the problem of the infinite was doomed to be falsified. In spite of Spinoza's great effort in this letter to provide a comprehensive taxonomy of infinities, it would turn out that one distinction which is paramount in the *Ethics* – the distinction between absolute infinity and infinity in its own kind – is completely absent in the letter.[47] A modest lesson one may take from this story is that when dealing with the infinite, one should not rush to announce having exhausted the issue.

---

[43] Ep. 54| IV/253/7-11. Italics added. A similar argument appears at the end of E1p17s, where Spinoza argues that insofar as God is the cause of both the essence and the existence of finite things, they must differ "and *cannot agree with [God] in anything except in name*" (II/63/30). Italics added. Cf. CM II 11| I/274/32-34.

[44] The claim that there is no *ratio* between the infinity of God and finite things also appears in Gersonides (*Milhamot ha-Shem* III, 3, 23b| *Wars of the Lord*, vol. II, 114.), quoted above, and in Crescas, *Or ha-Shem*, I, 3, 3. Cf. my "Crescas and Spinoza," 213–214.

[45] Ep. 12| IV/61/9-14.

[46] Ep. 12| IV/59/1-13 and IV/61/15-62/10. Spinoza's defense of actual infinity may provide further reasons for rejecting Bennett's reading. Clearly, in the *Ethics*, the absolute infinity of God is the strongest kind of infinity. Given Spinoza's admission of actual infinity, it would seem highly likely that God's infinity is actual (and actual infinity is clearly more than two). However, Spinoza does not explicitly use the terminology of actual infinity in the *Ethics*, and for this reason, I avoided resting my case on this consideration.

[47] The main reason for the absence of this distinction seems to be that at this stage Spinoza has not yet fully developed the distinction between the substance and its attributes. The notion of attribute is not at all mentioned in Ep. 12, and in Ep. 9 – written merely a month before Ep. 12 – Spinoza

In this paper, we have studied Spinoza's claim that God has infinitely many attributes. Against Bennett's interpretation of this claim as committing Spinoza to the existence of no more than two attributes, I have pointed out two strong philosophical reasons – as well as a theological tradition – that would have motivated Spinoza to hold that God has infinitely many attributes beyond thought and extension. We have exhibited numerous texts, both in the *Ethics* and outside it, in which Spinoza commits himself to the existence of infinitely many other attributes, and we have found not even a single text in which Spinoza asserts that God has – or even might have – only two attributes. We also explained Spinoza's flawless argument in Letters 64 and 66 concerning why human beings (and generally, the minds of modes of extension) cannot know any other attributes beyond thought and extension.[48] I submit that the case for taking Spinoza at his word, and reading "infinite attributes" as greater than any number, is as strong as it can be.[49]

Abbreviations

NS [*Nagelate Schriften*] refers to the text of the 1677 Dutch translation of Spinoza's works. Unless otherwise marked, all quotes from Spinoza's works and letters are from *The Collected Works of Spinoza*. 2 vols. Edited and translated by Edwin Curley. Princeton, NJ: Princeton University Press, 1985–2016. For the Latin text of Spinoza I have relied on: Spinoza, *Opera*. 4 volumes. Edited by Carl Gebhardt. Heidelberg: Carl Winter, 1925.

virtually identifies the notions of substance and attribute (see IV/46/20-23). For a study of the development of Spinoza's concepts of substance and attribute, see Melamed, "Glimpse into Spinoza's Metaphysical Laboratory."

[48] A reader who is still confused by Spinoza's arguments in Eps. 64 and 66 is invited to consult my detailed reconstruction of Spinoza's arguments in *Spinoza's Metaphysics*, 156–165.

[49] In passing, let me note that in his 1883 *Foundations of a General Theory of the Manifolds: A Mathematico-Philosophical Investigation into the Theory of the Infinite*, Cantor was engaged in a closed study of Spinoza's advocacy of actual infinity in Ep. 12, and of his theory of the infinite modes. Thus, for example, Cantor notes: "An especially difficult point in Spinoza's system is the relationship of the finite modes to the infinite one; it remains unexplained how and under what circumstances the finite can maintain its independence with respect to the finite, or the infinite with respect to still higher infinities" (Cantor, *Foundations*, 892). Cantor's discussion of the independence of finite modes clearly echoes the concerns raised by Hegel (for a discussion of the last issue, see my "Acosmism or Weak Individuals"). Overall, Cantor's discussion of the kinds of infinity in Spinoza is blended with his own views about the transfinite numbers. Thus, Bennett's mockery of "Spinoza and his contemporaries" who unlike Cantor "had just muddles and puzzles" (Bennett, *Study*, 76) seems somewhat out of place, as Cantor's writing seems to show that it was precisely "the muddles and puzzles" of Crescas, Spinoza, and Leibniz, that engaged Cantor and stimulated the development of his theory of transfinite numbers.

TdIE	Treatise on the Emendation of the Intellect [*Tractatus de Intellectus Emendatione*],
DPP	Descartes's Principles of Philosophy [*Renati des Cartes Principiorum Philosophiae Pars I & II*],
CM	Metaphysical Thoughts [*Cogitata Metaphysica*],
KV	Short Treatise on God, Man, and his Well-Being [*Korte Verhandeling van God de Mensch en deszelfs Welstand*],
TTP	Theological-Political Treatise [*Tractatus Theologico-Politicus*],
Ep.	*Letters*. Passages in the *Ethics* will be referred to by means of the following abbreviations: a(-xiom), c(-orollary), p(-roposition), s(-cholium) and app(-endix); 'd' stands for either 'definition' (when it appears immediately to the right of the part of the book), or 'demonstration' (in all other cases). Hence, E1d3 is the third definition of part 1 and E1p16d is the demonstration of proposition 16 of part 1.

Acknowledgements I would like to thank Zach Gartenberg for his helpful comments on an earlier version of this paper.

References

Ariew, R. (1990). The infinite in Spinoza's philosophy. In E. Curley & P.-F. Moreau (Eds.), *Spinoza: Issues and directions* (pp. 16–31). Leiden: Brill.

Bennett, J. (1984). *A study of Spinoza's "ethics"*. Indianapolis: Hackett.

Bennett, J. (1996). Spinoza's Metaphysics. In D. Garrett (Ed.), *The Cambridge companion to Spinoza* (pp. 61–88). Cambridge: Cambridge University Press.

Cantor, G. (1996). Foundations of a general theory of the manifolds: A mathematico-philosophical investigation into the theory of the infinite. In W. Ewald (Ed.), *From Kant to Hilbert* (Vol. II, pp. 878–919). Oxford: Clarendon Press.

Crescas, H. (1990). In R. S. Fisher (Ed.), *Or ha-Shem [light of the Lord]*. Jerusalem: Ramot.

Della Rocca, M. (1996). *Representation and the mind-body problem in Spinoza*. New York: Oxford University Press.

Descartes, R. (1964–1976). *Oeuvres de Descartes* [AT] (12 vols) (C. Adam, & P. Tannery, Eds.). Paris: J. Vrin.

Descartes, R. (1985). *The philosophical writings of Descartes* [CSM] (3 vols) (J. Cottingham, R. Stoothoff, & D. Murdoch, Trans.). Cambridge: Cambridge University Press.

Gersonides. (1560). *Milhamot ha-Shem*. Riva di Trento.

Gersonides. (1984–1999). *The wars of the Lord* (3 vols) (S. Feldman, Trans.). Philadelphia: Jewish Publication Society.

Harvey, W. Z. (2010). *Rabbi Hasdai Crescas* [Heb]. Jerusalem: Zalman Shazar Center.

Leibniz, G. W. (1923). *Sämtliche Schriften und Briefe*. Deutsche Akademie der Wissenschaften. Multiple vols. in 7 series. Berlin: Akademie Verlag, Cited by series, volume, and page.

Levy, T. (1987). *Figures de l'infini: Les mathématiques au miroir des cultures*. Paris: Éditions du seuil.

Melamed, Y. Y. (2010a). Acosmism or weak individuals? Hegel, Spinoza, and the reality of the finite. *Journal of the History of Philosophy, 44*, 77–92.

Melamed, Y. Y. (2010b). Spinoza's anti-humanism: An outline. In C. Fraenkel, D. Perinetti, & J. Smith (Eds.), *The rationalists* (pp. 147–166). Kluwer: New Synthese Historical Library.

Melamed, Y. Y. (2012). 'Omnis determinatio est negatio' – determination, negation and self-negation in Spinoza, Kant, and Hegel. In E. Förster & Y. Melamed (Eds.), *Spinoza and German idealism* (pp. 175–196). Cambridge: Cambridge University Press.

Melamed, Y. Y. (2013a). *Spinoza's metaphysics: Substance and thought*. Oxford: Oxford University Press.

Melamed, Y. Y. (2013b). Spinoza's metaphysics of thought: Parallelisms and the multifaceted structure of ideas. *Philosophy and Phenomenological Research, 86*, 636–683.

Melamed, Y. Y. (2014). Hasdai Crescas and Spinoza on actual infinity and the infinity of God's attributes. In S. Nadler (Ed.), *Spinoza and Jewish philosophy* (pp. 204–215). Cambridge: Cambridge University Press.

Melamed, Y. Y. (2015). A glimpse into Spinoza's metaphysical laboratory: The development of Spinoza's concepts of substance and attribute. In Y. Melamed (Ed.), *The young Spinoza* (pp. 272–286). Oxford: Oxford University Press.

Melamed, Y. Y. (2017). The building blocks of Spinoza's metaphysics: Substance, attributes, and modes. In M. D. Rocca (Ed.), *The Oxford handbook of Spinoza* (pp. 84–113). Oxford: Oxford University Press.

Melamed, Y. Y. (n.d.). Spinoza's metaphysics of substance. In: D. Garrett (Ed.), *The Cambridge companion to Spinoza* (2nd ed.). Cambridge: Cambridge University Press, forthcoming.

Pollock, F. (1966). *Spinoza: His life and philosophy* (2nd ed.). New York: American Scholar Publications.

Spinoza. (1925) *Opera* (4 vols) (C. Gebhardt, Ed.). Heidelberg: Carl Winter.

Spinoza. (1985–2016). *The collected works of Spinoza* (2 vols) (E. Curley, Ed., and Trans.). Princeton: Princeton University Press.

Chapter 6
Infinity in Spinoza's Therapy of the Passions

Sanja Särman

Abstract The ontological and epistemological priority of the infinite has been extensively dealt with in Spinoza scholarship. However, Spinoza's widely debated understanding of the infinite has not figured to the same extent in accounts of his therapy of the passions, the topic which this essay sets out to explore. My reasoning consists of six steps. First, I introduce Spinoza's cognitive therapy, which claims that we can be healed from our passions by acquiring adequate ideas of them; second, I show that Spinoza's cognitive therapy is subject to the Affective Force Condition, i.e., the claim that only a stronger affect can defeat another affect (4p7); third, I suggest that the one way an affect can be reliably weakened or strengthened is by us considering its intentional object as necessary or free, respectively; fourth, I argue that we can only adequately ascribe freedom to things if we regard them as absolutely infinite, and necessity to finite things if we can have an adequate idea of them as inscribed in an infinite causal nexus. This presupposes having an adequate idea of the infinite, and so fifth, I ask by which capacity, or from which perspective, we can acquire such an idea, on Spinoza's view. Sixth, I finally ask whether the infinity of the causal history of a finite thing qualifies as the kind of infinity which we can adequately conceive, and give a tentative (affirmative) answer to this question by situating this infinity in Spinoza's taxonomy of infinities.

6.1 Introduction

In contrast to an influential philosophical tradition represented by his predecessor Descartes, Spinoza maintained that "the human Mind has adequate knowledge [*cognitionem*] of God's eternal and infinite essence" (2p47).[1] In a key definition at

[1] When referring to the *Ethics*, I adopt the following shorthand, where (e.g.) 2p47 indicates the 47th proposition of the second part. app = appendix, a = axiom, c = corollary, d = demonstration (when

S. Särman (✉)
Philosophy Department, Hong Kong University, Hong Kong, China
e-mail: u3003474@hku.hk

O. Nachtomy, R. Winegar (eds.), *Infinity in Early Modern Philosophy*,
The New Synthese Historical Library 76,
https://doi.org/10.1007/978-3-319-94556-9_6

the outset of the *Ethics*, Spinoza defines God as "a being absolutely infinite, i.e., a substance consisting of an infinity of attributes, of which each one expresses an eternal and infinite essence" (1d6). Since God is essentially and not merely nominally defined in terms of his absolute infinity, knowledge of God necessarily involves knowledge of infinity. Spinoza therefore held that knowledge of infinity is possible for humans. Scholars have suggested different historical sources upon which Spinoza may have drawn in making this crucial move, and thereby have significantly clarified it.[2]

In this paper I will not continue their research, but instead discuss Spinoza's characteristic rehabilitation of the infinite as a possible object of knowledge in the context of his *cognitive therapy*, by which I will mean the "remedy for the affects which [...] consists in true cognition of the affects" (5p4s). We acquire "clear and distinct understanding" (5p4s) of our passions by considering what causes them, or their intentional objects,[3] in new light. This is so since, according to Spinoza's Causal Axiom, "the knowledge of an effect depends upon, and involves, the knowledge of the cause" (1a4). Only when we ascribe a different causal ancestry to our passion will we be able to conceive of the passion differently.

Edwin Curley has presented an account of how Spinoza intended this consideration of causes to function therapeutically. On his reading, Spinoza's "cognitive therapy" consists in reconsidering the intentional object of a passion in its causal context. The passion – for example, the hate I harbor toward any object X – is mitigated when I consider X to be caused by Y, which ideally redistributes some of my original hate toward X and makes Y fall victim to it instead. Curley admits that while this procedure may benefit my relationship with X, it may make it "harder for me to deal with Y in a constructive way" (1988, 133–4).[4] But thus conceiving of how the intentional object of each passion passes the buck to its cause does not seem to have any therapeutically liberating potential. For such regression in thought only benefits my relationship with X at the cost of disturbing my relationship with Y; it

following a proposition-indicating number) or definition (when following a Book-indicating number), l = lemma, p = proposition, s = scholium. For the English translation of Spinoza's works, I refer to Edwin Curley's editions (1988, 2016). For texts in Latin, I refer to Carl Gebhardt (1925).

[2] Ariew (1990) explores Spinoza's notion of the actually infinite in the light of fourteenth-century scholastic accounts of the categorematic infinite. Melamed (2014) does so in light of Hasdai Crescas' rejection of the Aristotelian argument against an actually infinite causal regress.

[3] For the sake of textual economy, the phrase "an idea of an affect/passion" can intend an idea of the intentional object of the affect/passion, if the affect/passion in question is intentional.

[4] Lin is critical of this model, which he states wrongly presupposes that any intentional affect "is a fixed quantity to be distributed among its causes" (2009, 281). Lin may be right; however, the example he supplies does not substantiate his case. According to Lin, if I learn that the poisoner of my dog had an accomplice, my original hate will not be divided between poisoner-1 and poisoner-2. Instead I will hate twice as much, since I will continue to hate poisoner-1, and come to hate poisoner-2 with the same intensity. Lin's example fails to illustrate his point since poisoner-1 and poisoner-2 co-cause the death of my dog. Instead, we ought to consider a case where poisoner-1's poisoning of my dog is caused by him being poisoned by poisoner-2 with a drug that causes one to compulsively poison dogs. In this case it does seem likely that at least some of my original hatred for poisoner-1 will be redistributed to poisoner-2.

can never even principally "destroy" (5p4s) the passion of which X was the initial object. It can only attenuate this passion indefinitely. In his treatment of Spinoza's therapy, Curley does not address the interpretative possibilities that arise when the causal ancestry behind the passion in question is seen as an actually infinite whole instead of as a potentially infinite regression.

By contrast, my thesis is that the notion of actual infinity is crucial for Spinoza's cognitive therapy, since the agent engaged in therapy should proceed by regarding her passions or their intentional objects as necessary, and she can do so only when considering them as infinitely caused. My goals are, accordingly, first, to show the importance of necessity and freedom in Spinoza's therapy of the affects. Second, I will indicate how the notions of necessity, freedom and infinity are linked, on Spinoza's view. Third, I will outline the distinctions Spinoza draws between different perspectives on infinity. Fourth, I will relate these perspectives *on* infinity to Spinoza's taxonomy *of* species of infinity in the *Ethics*. Hereby I will be able to briefly suggest one way in which we could conceive of the infinite causal ancestry of the finite modes that our passions intend. By completing these tasks, I hope I will be able to show the neglected significance the concept of infinity has in Spinoza's therapy.[5]

6.2 The Role of Necessity in Therapy

By 'affects' Spinoza means what we today call emotions. Spinoza's most fundamental distinction between kinds of affects is drawn between active affects and passions. Spinoza differentiates between active and passive affects in terms of the ideas that cause them: active affects "follow" from *adequate* ideas, whereas passions follow from *inadequate* ideas (3p3; see also 3p1c). Since affects, on this understanding, are characterized by the *ideas* that cause them, they are fundamentally cognitive; indeed, passions are even referred to as inadequate ideas *tout court* (Gen. Def. Aff. exp.).

We commonly lead lives dominated by the confused ideas of our passions – a condition which Spinoza calls *slavery*. To remedy this dismal condition, Spinoza, in the three last parts of his *Ethics,* formulates different strategies whereby our slavery to passions is to be undone. In 5p20s Spinoza lists five ways in which the Mind has "the power [...] over the affects". The first two both consist in knowledge of the affects, and read: "I. In the knowledge itself of the affects (see P4S); [and] II. In the fact that [the Mind] separates the affects from the thought of an external cause, which we imagine confusedly (see P2 and P4S)". Both I. and II. are founded on

[5] Contemporary studies of Spinoza's psychotherapy tend to neglect the notion of infinity. An exception is Ravven (2014) who addresses the existentially transformative aspect of acquiring a realization of the infinite network of causes in which agents are locked. While she compares Spinoza's philosophy to that of Maimonides, I aim to identify the condition of possibility for this realization in Spinoza's doctrine of the infinite.

5p4s, which implies that the empire of the passions can be swayed by us changing the way we *think* about our affects or their objects.[6] Since coming to *know* one's affect to be caused differently than one previously assumed is an instance of coming to know one's affect, I. and II. are both ways in which "clear and distinct understanding" of affects is deployed therapeutically.[7] In this essay I will refer to the activity whereby we cure ourselves from passions by acquiring adequate ideas and/or discarding inadequate ideas about these passions as Spinoza's *cognitive therapy,* or *therapy,* for short. I will not deal with the other strategies Spinoza mentions in 5p20 (which are not "derived" from 5p4).

One common charge against Spinoza's cognitive therapy is its intellectualism. However, it is rarely noted that this therapy is subject to the Affective Force Condition of 4p7.[8] 4p7 reads: "An affect cannot be restrained or taken away except by an affect opposite to, and stronger [*fortior*] than, the affect to be restrained" (see also 5p10d). This means that unless the adequate idea we form of a passion is accompanied by an affect "stronger" than the passion in question, the idea we have of the affect, no matter how adequate, will be inconsequential. Since ideas that cannot substantially transform our overall affective situation are therapeutically useless, we should strive to acquire ideas that are not only adequate but also affectively stronger than the passion to be outmaneuvered.

What, then, makes affects relatively strong? Spinoza maintains the "anti-intellectualist" (or more accurately anti-evidentalist) psychological thesis that the hierarchy of strength among affects is *not* established on the basis of the veracity of the constituent ideas of the affects in question. An affect is not stronger because its

---

[6] Spinoza's cognitive therapy has been widely criticized. Lin (2009, 271), who calls 5p3 "psychological alchemy" endorses Bennett's criticism (1984, 336) that since we cannot change the causal history of an affect, we cannot convert a passion to an action. For a recent defense of the general plausibility of 5p3 against this objection (as well as against criticisms from Della Rocca (2008) and Curley (1988)) see Marshall (2012).

[7] Throughout this essay I speak, as Spinoza himself does in the bulk of the *Ethics*, *as if* we can acquire adequate ideas. However, Michael Lebuffe (2010) has convincingly argued that since the deposit of adequate ideas is an eternal part of the mind, our adequate ideas cannot undergo any substantial change. Hence any improvement in our epistemic condition must consist in a decrease in our inadequate ideas in proportion to our adequate ideas. Although my interpretation is not incompatible with Lebuffe's I will follow Spinoza's manner of speaking in this essay.

[8] Bennett (1984, 332–3) notes that 4p7 implies that unless the realization that, say, jealousy is irrational is accompanied by an affect, say, of disgust and shame, the realization of the irrationality of the affect will be therapeutically inconsequential, but dismisses this finding for the following reasons: (1) it sounds "unSpinozistic"; (2) it contradicts Spinoza's dictum that "we shall determine by the mind's cognition alone the remedies of the affects" (5Pref). (1) is no argument. As for (2), the passage from 5Pref need not be read as stating that affects can be cured *only* by the mind's cognition – as this would contradict 4p7 and 5p10d – but could more charitably be read as stating that Spinoza will appreciate the efficaciousness of his therapy in mental and not physical terms. Similarly Lin (2009, 387) does not construe 4p7 as a condition imposed on any restraint of affects, but merely as one way that affects may be restrained. By contrast, Lebuffe's elucidation of "affective therapy" takes 4p7 very seriously (2010, 17). Interestingly, Nietzsche too seems to have observed, in a postcard to Overbeck from 1881, that in order for knowledge to be transformative for Spinoza, it has to become the most "powerful affect" (1967-).

constituent idea truthfully reports on what is evil or good (4p14d). While there may be many factors that determine the hierarchy of strength among affects, I will focus on the belief that Spinoza's metaphysical system seems designed to engineer, namely that finite things are *necessary*. In 5p5, Spinoza most explicitly presents the modality we attribute to the intentional object of our affects as the basis on which the hierarchy of strength among affects is erected:

> An affect toward a thing we imagine to be free is greater than that toward a necessary thing [*erga neccessariam* sc. *rem*] (by IIIP49), and consequently is still greater than that toward a thing we imagine as possible or contingent (by IVP11). But imagining a thing as free can be nothing but simply imagining it while we are ignorant of the causes by which it has been determined to act (by what we have shown in IIP35S). Therefore, an affect toward a thing we imagine simply is, other things equal, greater than that toward a necessary, possible, or contingent thing. Hence, it is greatest of all, q.e.d. (Translation modified.)[9]

Spinoza does not specify how we are to define freedom in order to be maximally affected by things we consider free, nor does he give a philosophical account of the necessity we should attribute to things in order to be minimally affected by them. This omission shows that whether we conceptualize the freedom of a thing along lines sanctioned by Spinoza's metaphysics (i.e. as self-causation: a freedom which God or Nature alone enjoys) or as the freedom he rejects (e.g. freedom of will) we will be equally affected by the object thus conceived, as long as we think of it "simply", without the mediation of thoughts of other things (see also 3p49d). Although freedom of will is on Spinoza's view impossible, attribution of such freedom strengthens the passions we have toward creatures we wrongly consider free in this way.[10] As a general rule, we can manipulate the differential strength of our affects by considering the object of the affect to be weakened as necessary and the object of the affect to be strengthened as free, regardless of whether we do so correctly or incorrectly.[11]

Notably, however, Spinoza insists that his cognitive therapy must proceed by "actions of the mind" (5p4s), and such actions are *adequate ideas*. 5p5 shows that one way in which this *cognitive strategy* for combatting passions can meet the Affective Force Condition is to proceed (i) by replacing inadequate ideas of free objects by ideas in which these objects are adequately considered as necessary; and/or (ii) by forming opposite ideas of alternative objects adequately conceived of as free.[12]

---

[9] In this passage Curley twice makes *neccessariam* the direct object of *imaginamur*. However, as 2p44 makes clear, we do not imagine things to be necessary, but understand them to be thus. Since it also not grammatically necessary to take *neccessariam* as the direct object of this verb (the accusative can depend solely on *erga*), I have modified the translation.

[10] The fatal consequences of our erroneous belief in freedom are described in 1app.

[11] I cannot here discuss the plausibility of this psychological claim in abstraction from Spinoza's overall philosophy. For a discussion, see Bennett (1984, 337–345).

[12] While Bennett (1984, 335–342) and Lin (2009, 270–72; 279–282) take the "knowledge of the affects" to be one strategy whereby reason can overcome the affects, and "understanding things as necessary" to be another, thus construing them as two different strategies, I interpret them as one. This would help explain why Spinoza, as Bennett (1984, 337) and Lin (2009, 270) put it, "oddly"

The criterion of adequacy inherent in Spinoza's cognitive therapy redirects our focus from Spinoza's account of the passions to his metaphysical story of the way things are. For if there is no agreement between the object and the way it is intended by the idea, the idea cannot be true and, therefore, not adequate either (see 2p34–2p35).[13] Let us next explore the relevant properties – free, contingent and necessary – in order to better understand how adequate understanding of them can be harnessed for Spinoza's therapeutic purposes.

6.3 The Role of Infinity in Necessity

In the world (as opposed to in its distorted mirror image in our passions) Spinoza maintains that things are either "necessary in reason of their cause" or "necessary in reason of their essence" (1p33s; see also CM, C. 306). I find it plausible that this disjunction is exhaustive.[14] If all things are necessary because of either reason, there seems to be no room for contingency.

In the *Ethics* Spinoza gives two alternative definitions of contingency, endorsing one and rejecting the other. At first, our belief in contingency – our belief that things could have been otherwise than they are – is seen as stemming from our ignorance of how the absolutely infinite substance renders things necessary (1p33s1). Later on (in 4def3 and 4def4), Spinoza distinguishes between contingency and possibility. In these passages, *contingency* belongs to everything whose essence does not involve existence, whereas *possibility* signifies what contingency did in 1p33s1 – the property of possibly having been otherwise, which we wrongly ascribe to things when we are ignorant of why they are necessary. Thus things that are not necessary in virtue of their essence but only in virtue of their causes can be contingent in this latter sense, even though they are necessitated by their causes. Let me address which protagonists of Spinoza's metaphysics assume these different roles.

---

passes over understanding things as necessary in his itemization of the power reason has over the passions in 5p20s. Colin Marshall follows Bennett and Lin in taking "understanding [things] as necessary" as a separate method for passion-control, irrelevant to the strategy of 5p3. He motivates this with the omission of 5p4 in the derivation of 5p6 where Spinoza first mentions the benefits of understanding things as necessary (2012, 144, n. 15). However, this omission cannot undermine the validity of the Affective Force Condition. Marshall's method for passion-control ("killing the mood") cannot be a reliable method, since it is likely that most passions will be accompanied by stronger affects than abstract reasoning about these passions. I propose adequately construing the affects to be weakened as necessary, and the objects of the ones to be strengthened as free, as a cognitive method which is more reliable than Marshall's, since it co-opts the affects for therapeutic purposes.

[13] Spinoza intensionally distinguishes between a true idea, defined in terms of agreement (in 1a6), and an adequate idea, defined in terms of its genesis in the mind alone (2p11c). However, he also maintains that the classes of adequate and true ideas are co-extensional (see Letter 60, C. 432).

[14] Don Garrett (1991) supplies ample textual evidence in support of the interpretation that the whole series of causal relations in the universe must be necessary on Spinoza's view.

First, there is the thing necessary in virtue of its essence or nature (Spinoza uses these terms interchangeably). This is God, Nature or substance (terms that in Spinoza's metaphysics co-refer), whose existence is necessary since it is "involved" in its essence (1d1). Since the existence of no other thing follows from its essence, everything else is contingent in this sense. Second, there are things which are necessary by reason of their causes.[15] As far as I can judge, things are necessary by reason of their cause, and God is uniquely free, *because* of the way God is necessary by reason of his nature.

On Spinoza's view, God is necessary by reason of his nature since this nature is *absolutely infinite* (1d6). As Gueroult has pointed out (1968, 172–3), the definition of God as absolutely infinite is intended to be genetic, that is, it should give the proximate cause of the thing in such a way that all its properties may be deduced from it.[16] From God's absolutely infinite nature, we are, supposedly, able to deduce all of his properties, which significantly includes freedom.

It is because God alone is absolutely infinite that he alone is free. By "free" Spinoza does not mean spontaneous or unconstrained, but constrained by one's nature alone (1d7). It is clear why freedom, thus conceived, follows from absolute infinity: absolute infinity implies that there *is* nothing that God is not. There is therefore nothing outside of God that could constrain him from without, and so he is free, in the Spinozist sense of being exclusively internally determined.[17]

Moreover, this absolutely infinite nature of God is identified with God's power (1p34). This means that there is no difference between what God necessarily is and what he necessarily produces. God both necessarily is, and necessarily causes himself to be, the absolutely infinite totality of all there is. This identification of power and essence is crucial to Spinoza's system. Nothing that is real can be denied of God, for this would deny him some reality – something that his absolute infinity does not allow (1p8s; Letter 35). Likewise nothing real can remain non-caused by him.

Crucially, reality, for Spinoza, comes in degrees. This much is clear from the statement that "the more reality or being a thing has, the more attributes belong to it" (1p9), for this would not be possible if there were no gradation of reality. Now, each thing that exists or is real, has a nature from which some effect follows (1p36). One is to judge which thing has less reality, and which more, on the basis of how

[15] God appears to be a curious member of this class, for he too is necessary in virtue of his cause – which he himself is (1d1). However, by "things necessary by reason of their cause" Spinoza presumably means "by reason of a cause *other* than themselves", which excludes God.

[16] In TdIE, the definition of a created thing is good when it first, gives the proximate cause of the definiendum, and second, is such that all the properties of the thing in question may be deduced from it. Spinoza gives the following example of such a definition: the circle "is the figure that is described by any line of which one end is fixed and the other is movable" (§ 98, C. 40).

[17] However, there is a spurious category of being – beings of reasons (*entia rationis*), such as universals (2p40s1), or aids of the imagination, such as numbers (Letter 12), which, since they are not properly speaking beings, fail to belong to God's nature. For Spinoza's claim that being is "badly divided into being and non-being" – and that beings of reason are thus wrongly termed beings, see CM, I:I.

causally efficacious the thing's nature is (see 1p9 and 1p16d).[18] This efficaciousness ranges from God, whose nature causes everything that is, to some finite thing that has only a minuscule effect on its surroundings. However, whatever has at least one effect, has a share in reality. Having some share in reality already qualifies something as a product of God's activity, which by definition produces all and only reality (1p34 and 1p8s together yield this result). It is therefore necessary that whatever has some reality, no matter how little, should be produced by God.

However, God cannot produce finite things directly. This is so since finite things, by definition, must "be limited by another of the same nature" (1d2). The limitation Spinoza speaks of is to be understood as causal (see Lin 2007, 282) or explanatory (for Spinoza, as Della Rocca [1996, 10–11] has argued, these coalesce). A thing is finite, in other words, if it is caused by, or can be explained by reference to, another thing that is like itself finite in kind, and so on *ad infinitum*. Hence it is clear that the very definition of finitude implies that all finite modes must coexist in mutual determination. If they are infinite – as they must be, given God's infinitely productive nature, their inter-determination will unfold infinitely.

The constraints the very definition of finitude places upon God's creation of finite things explains why Spinoza supplies us with two different metaphysical descriptions of the causal determination of any singular thing. On the one hand, 1p26 states that "[a] thing which has been determined to produce an effect has necessarily been determined in this way by God". If something produces an effect, it is real, and everything that is real belongs both to God's nature and to his range of causal efficacy, which, as per 1p34, come down to the same thing. This is so since God's absolute infinity implies that he is all there is – there is no reality which does not belong to God.

On the other hand, consider 1p28, where Spinoza claims that:

> No singular thing, or any thing which is finite and has a determinate existence, can neither exist nor be determined to produce an effect unless it is determined to exist and produce an effect by another cause, which is also finite and has a determinate existence; and again [...] to infinity.

On the one hand, the finite causal agent is determined by God to produce its effect; on the other hand, it is determined by its position in a causal nexus where finite things serve as links. It is only our, on Spinoza's view flawed, theological presumptions that make us assume that these two determinations must involve totally different and unrelated determinants. Spinoza's goal is to undermine these presumptions. For it is *through* – or more accurately, *insofar as it is* – every link in the infinite chain (described in 1p28) that the substance determines each finite thing causally (as Spinoza states it does in 1p26). In other words, God determines finite things by the means of an infinite mediation through their finite peers.

This should suffice to clarify what Spinoza means by claiming that everything is necessary either by reason of its essence or by reason of its cause. Moreover, the concept of infinity is crucial in rendering things free in the Spinozist sense and nec-

[18] For Spinoza's causally efficacious natures, see Valtteri Viljanen (2011), from whom I borrow the idea.

essary: while God is free in virtue of his *absolutely infinite nature,* finite things are necessary (by reason of their cause) in virtue of being inscribed in an *infinitely unfolding causal nexus*. However, this nexus itself is necessary because it is the only way God can produce an infinity of finite things, something which he, because of his absolutely infinite nature-cum-power, is necessitated to do. Let me now relate these insights to Spinoza's therapeutic project.

Recall that according to Spinoza's theory of the affects, the strength of an affect is positively proportional to the extent to which we consider the intentional object of the affect in question free. In order to be cured from our passions through cognitive therapy we should acquire adequate ideas whereby we conceive of finite things as necessary and discard inadequate ideas whereby we ascribe freedom to things other than what indeed possesses it – God.

On Spinoza's view, we can indeed come to know God as uniquely free if we refashion our concept of freedom as following uniquely from absolute infinity. Moreover, we can also come to think of things we previously held contingent as necessary. This transition is possible since no finite mode is necessarily *conceived* of as necessary. That this is Spinoza's position becomes particularly evident in Letter 12, where he distinguishes between two principal ways in which we can conceive of finite things. As long as we consider the essences of finite things in isolation from "the order of nature" (*ordo naturae*), Spinoza claims, "we can determine their existence and duration as we please, conceive it as greater or less and divide it into parts – without this doing any harm to our concept of them" (C. 202).

What does Spinoza mean by *the order of nature*? The phrase appears extensively in Spinoza's *oeuvre*. In 2p7s Spinoza provides "the whole order of nature" with the helpful gloss "the connection of causes". For present purposes, it suffices to note that according to 2p7, whenever a thing is conceived under the attribute of thought, its cause must also be so conceived, and so on to infinity. *The whole order of nature* thus signifies *the total succession of causes*. For finite modes, the succession of causal antecedents is infinite.

In Letter 12, Spinoza thus maintains that when we consider finite things in isolation from the infinite causal order in which they are inscribed, we are free to conduct thought experiments on their "existence and duration", neither of which we then consider as necessary. By contrast, when considered as part of the order of nature finite modes can "undergo none of these [thought operations] without our destroying at the same time the concept [*conceptum*] we have of them" (loc. cit. IV/55). When we (correctly: 1p28) consider finite things to be produced by an infinite array of causal connections among finite things, we cannot regard their "existence and duration" as being otherwise than they necessarily are. Only by taking into consideration the infinite causal history unfolding behind each finite mode can we adequately conceive of its existence as necessary (see 5p6d).[19]

[19] I agree with Newlands (2010) that necessity can only be ascribed to finite modes when the latter are considered in a maximally broad, indeed infinite, causal context, but find the claim that Spinoza is an anti-essentialist vis-à-vis the modality of these finite modes difficult to square with passages such as 2p44s.

This interpretation might give rise to the objection that by a *concept [conceptum]* Spinoza does not necessarily mean an *adequate idea,* or knowledge. However, in 2p44, Spinoza states that "[i]t is in the nature of Reason to regard things as necessary not as contingent". As is well known, reason is the second kind of knowledge (2p40s2:III), and the first kind of knowledge is, on Spinoza's view, "the only cause of falsity" (2p41). Since the sets of false and inadequate ideas are on Spinoza's view co-extensional, it follows that the idea we have of the existence of existing things as invariable and necessary is indeed adequate.

In Letter 12 Spinoza affirms that we are at least capable of possessing a concept of things as necessary in virtue of issuing from an infinite order of nature, namely the concept that is destroyed when we consider the same things in isolation from their causal history. Arguably,[20] then, we can know *that* a thing's causal ancestry is infinite without knowing each individual causal relation in the chain. Spinoza himself admits that he does not possess this latter knowledge – for it would amount to "knowing the whole of Nature and all of its parts" (Letter 32, C. II., 18).[21]

The possibility of knowing a finite mode's causal ancestry to be infinite without therefore knowing each of the causal antecedents is analogous to the way we, on Spinoza's view, can know an infinite God. For Spinoza's insistence that he can have as "clear an idea of God as [he has] of a triangle" – and thus, by 1d6, of a God consisting of infinite attributes, without therefore being able to browse through all these attributes (Letter 56, C. II, p. 423). Although Spinoza may not have endorsed this example, the case may be illustrated by the fact that one does not have to browse all natural numbers in order to establish the cardinality of the natural numbers, aleph-null. Since knowledge of the actual infinity *of* a series need not involve knowledge of each element *in* the series, Spinoza's postulation of knowledge of a thing as necessarily existing in virtue of being inscribed in an infinite causal nexus need *not* violate the assumption that knowledge of actually infinitely elements is infeasible for humans.

This assumption has recently figured in Della Rocca (1996, 183, n. 29), Marshall (2013) and prior to these, Diane Steinberg (1981), all of whom emphasize that the fact that the causal ancestry of each idea, *qua* a finite mode, is infinite is *prima facie* problematic for Spinoza's theory of adequacy. Marshall (2013, 22–3) summarizes the problem thus: "(a) Having an adequate idea of x requires having ideas of all x's causal antecedents; (b) Every finite mode has an infinite chain of causal antecedents; (c) No human mind can have ideas of an infinite chain of causal antecedents. […]; (d) Therefore, no human mind can have an adequate idea."[22] Both Steinberg (who submitted an account of Spinoza's third kind of knowledge as a solution to the

[20] The "exegetical" reason why one would want to harmonize what Spinoza says in Letter 12 with what he subsequently states in the *Ethics* is that he approved of the circulation of copies of the letter even after the completion of his magnum opus (see Melamed (2014, 205), who refers to Letter 81 for a proof of this).

[21] I thank an anonymous referee for highlighting this issue.

[22] Marshall does not defend (c), which he admits, "might be controversial" (2013, 22 n. 6); he only grants it for the sake for argument.

same problem prior to its articulation by Della Rocca) and Marshall argue that adequate knowledge is possible, since it is of eternal formal essences, elevated above the causal production of the finite things of which they are the essences (*pace* Kripke). Such knowledge need not involve infinite causal regresses.

By contrast, the problem of this paper is how Spinoza's cognitive therapy can be successful. My answer therefore concerns the *existence* of finite modes, not their formal and eternal essences. If someone is already contemplating the eternal essences of things instead of their messy generation and destruction, she is likely to be in little need of therapy. The subject of therapy needs a way to regard the object of her passion that does not *depart from* indifference for its existence. (Such indifference may be the endpoint of therapy, but it cannot plausibly be its starting point.) Although we cannot have adequate knowledge of the *existence* of finite things in many respects, we *can* know that it is necessarily infinitely caused. What kind of knowledge is this?

Spinoza maintains that, in the second kind of knowledge (2p40s2:III), we know properties common to all modes (2p38), as they follow from the nature of the attribute of which they are modes (2p13l). Accounts of the second kind of knowledge tend to focus on properties that follow uniquely from one attribute, and it is indeed by referring to properties common to all modes within an attribute (namely extension) that Spinoza intends to demonstrate 2p38. However, regardless of the attribute under which a mode is considered, "it follows from the nature of the attribute" (to borrow the phrase from 2p13l), qua infinite, that it is infinitely rich in modes.[23] To be necessitated by the means of an infinite causal nexus is a property common to all modes, under any attribute. Spinoza's assertion that the knowledge of things as necessary is of the second kind (2p44) presupposes that we can have adequate knowledge not only of properties common to attributes considered under an attribute, but of properties common to modes considered under *any* attribute.[24]

While knowledge of formal *essences* is clearly indispensable to the third kind of knowledge (2p40s), knowledge of the second kind of the necessary *existence* of finite modes allows us to weaken the reign of the passions by reason. And it has transpired that in order to regard finite things as necessary, we must see them as locked in an infinite network of causes. Attributions of necessity thus depend on the vertiginous condition that humans have the capacity to posit the infinity of the causal ancestry that renders a finite thing necessary. Let me now address the question of how we, on Spinoza's view, should conceive of the infinite causal ancestry

[23] Della Rocca has argued that there could be no sufficient reason for an infinitely creative attribute to create less than an infinity of finite modes (2008, 71).

[24] If common notions within an attribute are scientific in nature (belonging to physics in the case of Extension, to science of mind in the case of Thought), then notions common to all modes insofar as they are modes of any attribute at all, are ontological. Just as transcendental terms were applied "transcategorically" in late Scholastic sciences of being, predicates such as *being infinitely caused* apply to modes across attributes. Indeed, without attribute-neutral knowledge of what is true of a mode insofar as it is a mode, the Spinozist edifice crumbles. For propositions such as 1p28 are not attribute-specific, yet purport to supply us with knowledge of how finite modes are caused.

of each passion (or its intentional object) so as to be able to regard it adequately as necessary.

6.4 Perspectives on Infinity. The Twofold Distinction

The two key texts where Spinoza discusses the capacity whereby we acquire adequate understanding of infinities are 1p15s and Letter 12 of July 1663, otherwise known as "The Letter on the Infinite", addressed to Spinoza's friend Lodewijk Meyer. In both these texts, Spinoza distinguishes the divisible infinity that we imagine from the indivisible infinity that the intellect alone can conceive.

In 1p15s (as well as in Letter 12) Spinoza concedes that extension (in Letter 12, "extended substance") cannot be considered as an infinitely divisible quantity without risking paradoxes *à la Zenon*. This, however, does not disprove that extension is indeed an attribute of God's and as such infinite. Spinoza merely takes such difficulties to suggest that extension should not be thought of as divisible into or composed of parts. Instead, extension is an infinite and indivisible quantity. The intuitive objection to Spinoza's claim is that while a quantity may of course be *physically* indivisible, any quantity seems to be *conceptually* divisible into lesser constituent parts. However, Spinoza denies that our human incapacity to imagine an indivisible quantity can be adduced as a ground against the existence of such a quantity. Spinoza maintains that our propensity for conceiving of quantity as divisible stems from our *imagination* (1p15s[V]), the source of error. In 1p15s there are thus two principal ways in which we conceive of an infinite quantity: either with the intellect as an indivisible whole or with the imagination as potentially infinitely divisible.

This distinction recurs in Letter 12. Spinoza's point in this letter is subtly different, as he here contends that we will successfully make the overarching distinction between infinities "that cannot be divided into any parts" and infinities that "can be divided into parts without any contradiction" (Letter 12, C. 201), on the condition that we first make three preliminary distinctions.[25] One of these preliminary distinctions is nonetheless again drawn between the infinity which "we can only understand but not imagine and [the infinity which] we can also imagine" (Letter 12, C. 201). Significantly, the kind of infinities which the imagination can conceive are the ones that can be mathematically manipulated – Spinoza uses the dichotomy series as an example, and, in the vein of the Aristotelian tradition, construes it as a potential infinity (Letter 12, C. 203).

In both these texts, Spinoza maintains that we can form adequate ideas of actually infinite wholes by the means of our intellect – even if we only succeed in this enterprise rarely, and with great difficulty (1p15s [V]; Letter 12, C. 202–3). But the *imagination* can only inadequately conceive of existing infinities. It is interesting,

[25] The other distinctions are causal and mathematical, respectively (Letter 12, C. 201). Concerning the former distinction I agree with Gueroult (1966, 387–90) that whereas the substance is infinite "by the force of its definition", finite modes can be called infinite "by the force of [their] cause".

and certainly not arbitrary, that on Spinoza's view *necessity* and *contingency* are also assigned separately by these capacities. For Spinoza holds that it "it depends only on the imagination that we regard things as contingent" (2p44c1).

Combining the view that the intellect alone can adequately conceive of actual infinity with the view that necessity can only be adequately ascribed to finite things if we consider them as locked in an infinite succession of causes, it seems plausible that it is because of the imagination's incapacity to adequately conceive of actual infinity, that it can only see things as contingent, but not as necessary. By considering the causal ancestry of a finite thing as a potentially infinite succession of causes, we can never realize its necessity. For we are then conceptualizing the succession temporally or numerically – but time and number are merely products of the imagination (2p44s; Letter 12, C. 204).

It is the actual infinity *a parte ante* of finite causes behind our passions that we are to conceive intellectually and cease to imagine, if we are to adequately regard finite modes as necessary. In order to suit Spinoza's therapeutic regimen, we must be able to first, in a pre-healed state, imagine this infinity, and then cease to imagine it and conceive it with the intellect alone. In order to understand which infinity can be both imagined and understood, and whether the infinite causal nexus of finite things qualifies as this kind of infinity, let us now turn to the different infinities Spinoza subscribes to in the *Ethics*. By discussing this taxonomy of infinities, I hope I will be able to suggest one way in which we can form an adequate idea of the infinite ancestry of our passions.

6.5 Species of Infinity. The Threefold Distinction

Above I outlined the two perspectives from which we can consider infinities, on Spinoza's view. Independently of how we conceive them; however, Spinoza asserts that there *are* three different species of infinities instantiated in the world.

In 1p16dem Spinoza summarizes the three infinities that exist in this way: "But since the divine nature has *absolutely infinite attributes* (by D6), each of which also expresses an essence *infinite in its own kind*, from its necessity there must follow *infinitely many things* in infinite ways (i.e., everything which can fall under an infinite intellect), q.e.d" (Emphases added; Translation modified). Here I will call these infinities absolute, qualitative and quantifiable, respectively.[26] In this paper I will not deal with all the intriguing difficulties this division poses, but merely state how these infinities relate to the perspectival distinction mentioned above, with the purpose of further elucidating Spinoza's therapy.

[26]Following Leibniz (2001, 43), who, in a note on Spinoza's *Ethics*, dating from around 1676, similarly distinguishes between three infinities. The infinities Leibniz distinguishes between are: "mere" infinity (*tantum infinitum*); infinity which is "greatest in its own kind" (*maximum in suo [...] genere*) and infinity which is "everything".

First, there is absolute infinity. This is the kind of infinity which belongs only to that whose essence involves existence, God or Nature. Spinoza maintains that God is such that we cannot imagine him, but only understand him with the help of the intellect (Letter 12, C. 203; Letter 56, C. II., 423). The potential infinity the imagination conceives by numerical measures therefore fails to represent the infinity of Nature.

Second, there is *perfection* or *infinity in kind*, which I propose to call qualitative infinity.[27] Whereas things finite in kind are limited by other things that are like them finite in kind – the example Spinoza adduces in 1d2 is that of a body, for another body can always be conceived as greater than any body – things that are infinite in kind are not so limited. The most paradigmatic example of something that is qualitatively infinite are God's attributes, notoriously defined as "what the intellect perceives as constituting the essence of a substance" (1d4). Suffice it here to state that since attributes are what we conceive of as constituting the essence of a substance, and since this essence must be self-sufficient, attributes cannot be conceived in terms of something other than themselves (1p10). Since composition involves some explanatory dependence, where the compound is explained by reference to its parts, attributes cannot be conceived of as composed of parts, or as divisible. On Spinoza's view there is therefore no danger that qualitative infinity should be divisible into parts. Qualitative infinity comes, as it were, woven in one piece. Since only the intellect can conceive of indivisible and infinite wholes, only the intellect can conceive of qualitative infinity.

Third, there is "quantifiable infinity". By this I mean the kind of infinity that we *can* quantify. Outside of the abstract realm of mathematics, this quantification is illicit. For the imagination necessarily employs aids (*auxilia*) – number, measure and time – as it conceives of classes of modes, quantity and duration, respectively (Letter 12, C. 203, G. IV/57), tools that themselves have no reality outside of the imagination.[28] Since modes, quantity and duration can all be quantified, the infinity of each of these may be called 'quantifiable'. Clearly, the infinite nexus of finite modes causing each other belongs to this category, for they constitute one class of modes. Although there is an infinity we can understand but not imagine, there is

[27] In Letter 36 to Hudde (June, 1666), Spinoza draws on the difference between infinity in kind and absolute infinity to illuminate his cosmological proof of God's existence.

[28] It seems Spinoza considered infinities that in his day were mathematically manipulated as infinite only in name. This is indicated by Spinoza's mathematical distinction in Letter 12 between "what is *called infinite* [*infinitum dicitur*] because it has no limits and that whose parts we cannot explain or equate [NS: determine or express] with any number, though we know the maximum and minimum" (C. 201; G. IV/53: Emphasis added). Toward the end of Letter 12, Spinoza again indicates that infinity is duly predicated both of what is infinite in virtue of its nature and what is infinite in virtue of its cause, while "others" – presumably mathematical infinities "are *called infinite* – or if you prefer, indefinite – because they cannot be equated with any number though they can be conceived to be greater or lesser" (C. 205: Emphasis added). These two instances seem to imply that for Spinoza the infinities we construct mathematically are nominally infinite, but not real infinities.

none that we cannot understand (Letter 12, C. 201), and so we can conceive quantifiable infinity with our intellect as well as imagine it.

This overview of the relation between perspectives on infinity and species of infinity can be summarized as follows: out of the three infinities that exist – whether they be instantiated in modes, attributes or substance -, only quantifiable infinity can be imagined. All infinities can be conceived by the intellect, or, which is the same, we can have adequate ideas of all these infinities. It should be emphasized that this reading is not without difficulties – but these difficulties are, as I believe, internal to Spinoza's own texts.

One widely debated difficulty concerns the infinity of the attributes. Although each attribute is as such qualitatively infinite (expressed more clearly, this property belongs distributively to the class of attributes), collectively the attributes are infinitely *many*: "By God I understand a being absolutely infinite, i.e., a substance consisting *of an infinity of attributes*, of which each one expresses an eternal and infinite essence" (1d6: Emphasis added).[29]

The difficulty is this: if each attribute is, as Spinoza implies in Ep. 9 (C. 195–6), a nature that the intellect correctly attributes to the substance, then how can something be asserted about the attributes – namely, that they are infinitely many – which does not pertain to the substance, which is unique? This seems to sin against the transitivity of identity.[30] I cannot resolve this difficulty here. Instead, I will address the limited issue of how absolute infinity in the substance can imply a quantifiable infinity of attributes. This will indicate one way in which the intellect can conceptualize quantifiable infinity without relying on the imagination.

Although it does not follow from the claim that the substance possesses *all* the attributes that it possesses an *actually infinite amount of* attributes, it would follow from the claim that it possesses them perfectly, in the sense of unsurpassably or qualitatively infinitely. This interpretation is supported not only by passages in CM[31] where Spinoza's statements about perfection could perhaps be brushed aside

[29] Bennett (1984, 75–9) proposes that the word 'infinite' could instead be construed as 'all', and that nothing of importance in Spinoza's system demand that all the attributes should be more numerous than the two we know (extension and thought). This reading has been disputed by Ariew (1990) in response, not to Bennett, but to the same position as articulated by Abraham Wolf (1926, 190–1) and Kline (1977, 341–7), and by Melamed (most recently in forthcoming, see also Melamed [2013, 166–167; 198–199]), in response to Bennett. As Bennett himself admitted, that God should possess only two attributes is incompatible with passages such as 2p7s, Letters 63 and 65. Bennett formulates his interpretation not to explicate what Spinoza says, but to grant him only what he is obliged to say in order for his system to hold strong, and on Bennett's view the infinity of the attributes is so far from being a stronghold of the system that "in the *Ethics* it is idle" (1984, 79). Since my present purpose is to explore whether and how Spinoza can consistently grant us knowledge of the infinite ancestry of our passions, *given* his philosophical system as conceived in the *Ethics*, I grant that the set of attributes is, on Spinoza's view, infinite.

[30] That is, presupposing the objectivist reading of the attributes, which dominates contemporary Spinoza scholarship. For a helpful overview as well as criticism of the subjectivist/objectivist division in Spinoza studies, see Shein (2009).

[31] See for instance CM II:III (C. 319): "So only God is to be called absolutely infinite, insofar as we find that he really consists of infinite perfection."

as expositions of positions taken by Scholastic predecessors, but also by a passage in the *Ethics* where Spinoza speaks of absolute infinity in terms of ultimate perfection (1p11s; see also 2p34).

On my reading, the substance is absolutely infinite since it possesses the qualitatively infinite attributes in a *qualitatively infinite way,* that is, so that no possession of attributes could be greater or more perfect than that of the substance. Metaphorically put, absolute infinity can be understood as qualitative infinity raised to adverbial power.[32] Arguably, if the substance just possessed all the attributes, and they amounted to any arbitrary number, it would not possess them in an absolute way, but in the very measurable and therefore surpassable way that Spinoza insists cannot pertain to the substance.[33] Therefore it must possess them perfectly, and they must, as a consequence, be actually infinitely many.

I suggest that conceiving of the quantifiable infinity of attributes *as a perfection of the substance* is one way in which the intellect can conceive of a quantifiable infinity without resorting to the tools of the imagination. For perfection, or qualitative infinity, we conceive with the intellect alone. If we can indeed conceive of the quantifiable infinity of attributes as a perfection of the substance, it seems we could similarly conceive of the infinity of causally related finite modes. The causal ancestry of each finite mode intended by a passion is infinite, since the web in which it is inscribed is a maximal manifestation of a perfectively productive God. In spite of the etymology of the word, this "perfection" is not teleological; it refers only to the infinity of God's nature, of which Spinoza insists we can have an adequate idea.[34]

Let me briefly sum up this count of infinities. There are three different ways in which things are infinite in Spinoza's metaphysics: the modes – which are infinitely many – are quantifiably infinite, the attributes are both quantifiably and qualitatively infinite (collectively and distributively, respectively), and the substance is absolutely infinite, in the sense that it perfectly possesses the attributes. Out of these, only quantifiable infinity is imaginable. We habitually construe the quantifiably infinite mathematically, as potentially infinite successions (in Letter 12, C. 203, Spinoza uses the division of time as an example). Yet potential infinity does not, on Spinoza's view, correctly represent either the infinity of extension or the collective multiplicity of the attributes. Nor can we conceive of the causal ancestry of our passions as a potential infinity, if we want to regard finite things as necessary. On the contrary, to

[32] While *absolute* in the phrase *natura divina infinita absolute attributa habeat* qualifies the infinity of the attributes, not that of the divine nature (stylistically, the figure is a *transgressio*: I thank Josef Eskhult for discussion of the position of the adverb), I agree with Kline (1977, 344) that the adverb collectively characterizes the attributes *when possessed by the substance*.

[33] Deleuze (1990, 35) also invokes the reasons stated in 1p15s, stating that: "[W]e cannot pass through just three or four attributes without bringing back into the absolute the same numerical distinction which we have just excluded from infinity".

[34] Spinoza uses 'perfection' in two senses – as a teleological notion established on subjective and anthropocentric standards in 4app, and in an "objective" ontological sense, as synonymous with reality or being (see 1p11s; 1p16d and 2d6). This latter a-teleological (but theological) perfection is attributed to the world, which clearly encompasses the order of finite modes, in 1p33s2 and Letter 36.

successfully combat our passions by realizing their necessity we should think of their causal ancestry in the intellect's terms – as an actually infinite whole rather than as potential infinity.

6.6 Concluding Remarks

In this paper I have given the infinite in Spinoza's philosophy a therapeutic twist. The cognitive therapy which is to cure us from our passions can succeed if we can think adequately of the infinite causal nexus in which the finite objects of our passions are situated. Of course, we cannot have adequate knowledge of each of the individual causes behind any finite mode, for these are not common to all. Nonetheless, Spinoza insists, we can have adequate knowledge of the infinity of the series, in the sense that we can know the proximate cause of the whole of the series: God's absolutely infinite nature. But how is this possible? In this paper, I have suggested that we can conceptualize the infinite causal ancestry of our passions as one of God's perfections. On this reading, I can come to regard the object of my passion as part of nature more easily, to the extent that I accept the traditional definition of God as a "most perfect being" (to which Spinoza certainly subscribed, see e.g. KV II, 18; CM II, 2, Letters 2 and 60, and 1p11d). For if I consider the whole infinite nexus of causes behind my passion as one way in which God's necessarily infinite power-cum-essence manifests itself, I will be less inclined to terminate the blame arbitrarily at the object of my passion, or its cause, or the cause of its cause, and so on ad infinitum.

By considering God, and not any seemingly contingent thing, as the ultimate cause of my sadness, I, in effect, "separate the affect" from the partially inadequate idea of infinite "external causes" (5p2d) and instead join the affect to a "true thought", and also, additionally, to the true thought of an object whose causes do not proliferate infinitely, namely the *causa sui*, or God. Spinoza emphasizes the value of thus adequately blaming God for the passions we necessarily undergo in 5p18s: "as we understand the causes of Sadness, it ceases (by P3) to be Sadness. And so, insofar as we understand God to be the cause of Sadness, we rejoice".

I have suggested that it is only by having an adequate concept of the actual infinity of the causal series unfolding behind whatever passion we suffer from that we come to regard it as necessary and God as its cause. But for this reason, adequate conceptualization of the actually infinite is also the *sine qua non* of love of God (*amor Dei*). For in order to love God, we must be able to conceive of him as a cause (Def. Aff. VI; 5p32s). But as long as we insist on thinking infinity through the imagination we are unable to see how God causes any finite mode. For then we can only pass along a seemingly contingent, potentially infinitely unfolding causal narrative, but never realize the infinite whole that he, *qua* all, freely causes, and by the means of which our passions are necessitated. Unless we had a different perspective on infinity to have recourse to, we would not be able to ascend from contingency – in the sense of inadequately understood causal chains – to the full insight of God's freedom as instantiated in our slavery. On Spinoza view, there *is* a perspective that admits of this insight. But because it is so psychologically hard to acquire and sustain this perspective (1p15s [V]; Letter 12, C. 202–3), such insights are bound to be both rare and difficult, and perhaps, we might add, frustratingly transitory. And yet such insights might be all the freedom a human being can have.

Abbreviations

AT *OEuvres de Descartes*, éd. par Ch. Adam et P. Tannery, nouv. présent. par J. Beaude, P. Costabel, A. Gabbey et B. Rochot, 11 vols., Paris, Vrin 1964–1974.

CSM *The Philosophical Writings of Descartes*, ed. and transl. by John Cottingham, Robert Stoothoff, Dugald Murdoch, vols. 1 and 2. Cambridge, University Press, 1984–5.

Ethics *The Collected Works of Spinoza*, ed. E. M. Curley. Princeton: Princeton University Press, 1985. When referring to the *Ethics*, I note the part of Ethics followed by A= axiom, cor.= corollary, dem.= demonstration, P= proposition, or Schol.= scholium, with their respective numeration, e.g. "2P47" refers to Part Two of the Ethics, Proposition 47.

References

Ariew, R. (1990). The infinite in Spinoza's philosophy. In E. Curley & P.-F. Moreau (Eds.), *Spinoza, issues and directions – The proceedings of the Chicago Spinoza conference* (pp. 16–31). Leiden: Brill.

Bennett, J. (1984). *A study of Spinoza's ethics*. Indianapolis: Hackett Publishing Company.

Curley, E. (1988). *Behind the geometrical method – A reading of Spinoza's ethics*. Princeton: Princeton University Press.

de Spinoza, B. (1925). In C. Gebhardt (Ed.), *Opera* (Vol. 1–4). Heidelberg: Carl Winter.

de Spinoza, B. (1988, 2016). *The collected works of Spinoza* (Vol. 1–2, E. Curley Ed. and Trans.). Princeton: Princeton University Press.

Deleuze, G. (1990). *Expressionism in philosophy: Spinoza* (M. Joughin, Trans.). New York: MIT Press.

Garrett, D. (1991). Spinoza's necessitarianism. In Y. Yovel (Ed.), *God and nature: Spinoza's metaphysics* (pp. 191–218). Leiden: Brill.

Gueroult, M. (1966). La lettre de Spinoza sur l'infini (Lettre XII, à Louis Meyer). *Revue de Métaphysique et Morale, 71*(4), 385–411.

Gueroult, M. (1968). *Spinoza, vol. I: Dieu*. Paris: Aubier-Montagne.

Kline, G. (1977). On the infinity of Spinoza's attributes. In S. Hessing (Ed.), *Speculum Spinozanum, 1677–1977* (pp. 333–352). London: Routledge and Kegan Paul.

Lebuffe, M. (2010). *From bondage to freedom: Spinoza on human excellence*. Oxford: Oxford University Press.

Leibniz, G. W. (2001). On Spinoza's ethics and on the infinite. In *The labyrinth of the continuum, writings on the continuum problem* (R. T. W. Arthur, Ed. and Trans.). Yale: Yale University Press.

Lin, M. (2007). Spinoza's arguments for the existence of God. *Philosophy and Phenomenological Research, 75*(2), 269–297.

Lin, M. (2009). The power of reason in Spinoza. In O. Koistinen (Ed.), *The Cambridge companion to Spinoza's ethics* (pp. 258–283). Cambridge: Cambridge University Press.

Marshall, C. (2012). Spinoza on destroying passions with reason. *Philosophy and Phenomenological Research, 85*(1), 139–160.

Marshall, E. (2013). *The spiritual automaton*. Oxford: Oxford University Press.

Melamed, Y. (2013). *Spinoza's metaphysics*. Oxford: Oxford University Press.

Melamed, Y. (2014). Hasdai Crescas and Spinoza on actual infinity and the infinity of God's attributes. In S. Nadler (Ed.), *Spinoza and medieval Jewish philosophy* (pp. 204–215). Cambridge: Cambridge University Press.

Melamed, Y. (forthcoming). Spinoza's metaphysics of substance. In D. Garrett (Ed.), *The Cambridge companion to Spinoza* (2nd ed.). Cambridge: Cambridge University Press.

Michael, D. R. (1996). *Representation and the mind-body problem in Spinoza*. New York: Oxford University Press.

Newlands, S. (2010). The harmony of Spinoza and Leibniz. *Philosophy and Phenomenological Research, 81*(1), 64–104.

Nietzsche, F. (1967). Letter 135. In G. Colli & M. Montinari (Eds.), *Digitale Kritische Gesamtausgabe*. Berlin: de Gruyter.

Ravven, H. M. (2014). Moral agency without free will: Spinoza's naturalizing of moral psychology in a Maimonidean key. In S. Nadler (Ed.), *Spinoza and medieval Jewish philosophy* (pp. 128–151). Cambridge: Cambridge University Press.

Rocca, D. (2008). *Spinoza*. New York: Routledge.

Shein, N. (2009). The false dichotomy between the objective and subjective interpretations of Spinoza's attributes. *British Journal for the History of Philosophy, 17*(3), 505–532.

Steinberg, D. (1981). Spinoza's theory of the eternity of the mind. *Canadian Journal of Philosophy, 11*(1), 35–68.

Viljanen, V. (2011). *Spinoza's geometry of power*. Cambridge: Cambridge University Press.

Wolf, A. (1926). Spinoza's conception of the attributes of substance. *Proceedings of the Aristotelian Society, 27*, 177–192.

Chapter 7
The Road to Finite Modes in Spinoza's Ethics

Noa Shein

Abstract There are many aspects of the *Ethics* that seem to suggest, or perhaps even require the possibility of deducing finite modes from the infinite substance. Nonetheless, as many have noted even during Spinoza's own time, it is far from clear that such a deduction can be successfully performed. In this chapter I argue that the expectation of a top-down deduction (i.e., one that begins with an attribute and terminates with a finite mode) is unwarranted, and that interestingly enough, it is not only unwarranted with regard to Spinoza but with regard to Descartes as well. I show this by pointing to the crucial role confusion plays for both, noting that our epistemic journey to clear and distinct or adequate knowledge begins with confusion, and that this trajectory is one of emending initially confused ideas. This shows that epistemically a bottom-up trajectory necessarily precedes a top-down one. My claim, however, goes beyond this point regarding the order of discovery. I argue that the state of confusion presupposes a plurality of finite modes, in Spinoza's case, and substances and modes in Descartes's. Confusion, I claim, amounts to an inadequate perception of an amalgam of ideas of a real plurality of things. Recognizing the metaphysical ground of confusion along with its central epistemic role brings to light the artificiality, and indeed ultimately the impossibility of a top-down deduction which is divorced from a preceding bottom-up analysis.

I would like to thank Alan Nelson, Julie Klein and Michael Della Rocca for their helpful comments as well as the participants of the "Infinity in Early Modern Philosophy" workshop, held in Jerusalem in 2016, for their discussion. The research for this paper was made possible by the Israel Science Foundation (grant no. 1199/13).

N. Shein (✉)
Philosophy Department, Ben-Gurion University of the Negev, Beer-Sheva, Israel
e-mail: sheinn@bgu.ac.il

O. Nachtomy, R. Winegar (eds.), *Infinity in Early Modern Philosophy*,
The New Synthese Historical Library 76,
https://doi.org/10.1007/978-3-319-94556-9_7

7.1 Introduction

There are many aspects of the *Ethics* that seem to suggest the possibility of deducing finite modes from the infinite substance. Among these are Spinoza's axiom that "knowledge of an effect depends on and involves the knowledge of the cause" (1A4), 1P16: "From the necessity of the divine nature there must follow infinitely many things in infinitely many modes, (i.e., everything which can fall under an infinite intellect)," his insistence in 2P10Schol that the correct order of philosophizing begins with a contemplation of God's nature, his characterization of intuitive knowledge as proceeding from the adequate knowledge of the formal essence of an attribute to things, the very order of the first two parts of the *Ethics*—beginning with "Of God" and only then turning to "of the Mind", and the geometric exposition itself also seems to elicit this type of expectation.[1]

Considering these texts, it seems it should be possible to begin with an adequate idea of the formal essence of one of the attributes, say Extension, and deduce from it ideas of particular bodies or finite modes, or better yet, see how they follow from infinite Extension. Many have taken the relevant deduction here to be that of logical entailment.[2] This way of proceeding would track the ontological order, going from cause to effect, and from what is in itself to what is in another. In spite of the inviting and elegant nature of this top-down trajectory, once we try to actually follow it, things become complicated very quickly. Spinoza's contemporary Tschirnhaus noted the difficulty, as have Hegel and our contemporaries Curley and Della Rocca among others.[3] These commentators identify what seems to be an unbridgeable gap

[1] All references to the *Ethics* are from Benedictus de Spinoza, *The Collected Works of Spinoza*, ed. E. M. Curley (Princeton: Princeton University Press, 1985). References to Spinoza's letters are from Benedictus de Spinoza et al., *The Letters* (Indianapolis: Hackett Pub. Co., 1995). I have used the following abbreviations to refer to Spinoza's writings: Ep. = letter followed by the standard numeration. When referring to the *Ethics*, I note the part of *Ethics* followed by A= axiom, cor.= corollary, dem.= demonstration, P= proposition, or Schol.= scholium, with their respective numeration, e.g. "2P47" refers to Part Two of the *Ethics*, Proposition 47.

[2] See for example Stephen Nadler, "Spinoza's Monism and the Reality of the Finite," in *Spinoza on Monism*, ed. Goff Philip, Philosophers in Depth (UK: Palgrave Macmillan, 2012). Yitzhak Melamed, "Why Is Spinoza Not an Eleatic Monist (or Why Diversity Exists)," ibid., ed. Philip Goff (London: Palgrave). Karolina Hübner, "Spinoza on Negation, Mind-Dependence and the Reality of the Finite," in *The Young Spinoza: a Metaphysician in the Making*, edited by Yitzhak Melamed (New York: Oxford University Press, 2015), 221–22. Samuel Newlands, "Another Kind of Spinozistic Monism," *Noûs* 44, no. 3 (2010). Martin Lin, "Substance, Attribute, and Mode in Spinoza," *Philosophy Compass* 1, no. 2 (2006). E. M. Curley, *Behind the Geometrical Method: A Reading of Spinoza's Ethics* (Princeton: Princeton University Press, 1988), 38–39.

[3] Georg Wilhelm Friedrich Hegel and George Di Giovanni, *The Science of Logic* (Cambridge: Cambridge University Press, 2010), 328. E. M. Curley and Gregory Walski, "Spinoza's Necessitarianism Reconsidered," in *New Essays on the Rationalists*, ed. Rocco J. Gennaro and Charles Huenemann (New York: Oxford University Press, 1999). Michael Della Rocca, "Striving, Oomph, and Intelligibility in Spinoza," in *Judgement and the Epistemic Foundation of Logic*, ed. Maria Van der Schaar (Springer Science & Business Media, 2013). "Steps toward Eleaticism in Spinoza's Philosophy of Action," in *Freedom and the Passions in Spinoza's Ethics*, ed. Noa Naaman Zauderer and Tom Vinci (Cambridge University Press, Forthcoming). For an analysis of

between what is absolutely infinite, full of being and wholly undifferentiated and what is "in part a negation", differentiated and finite. This kind of worry has driven some to hold that finite modes aren't fully real for Spinoza—Hegel being the most salient example of course. Della Rocca has been advocating as well for an ideal status for modes (both finite and infinite). The way he phrases the issue is that no bifurcation in nature can withstand the principle of sufficient reason.[4] Here the issue is that the deduction that Della Rocca seeks is one that can be grounded in sufficient reason, but, according to him, no differentiation can be so grounded. Curley and Walski face the problem by arguing that the necessity that attaches to finite modes ought to be understood as "moderate" rather than "strict" in order to avoid accepting the non-reality of finite modes.[5] Curley and Walski highlight the issue by noting that laws of nature require antecedent conditions to deduce particular events. Since the antecedent conditions cannot be given *a priori*, so to speak, they advocate for a moderate necessitarianism allowing for different "possible worlds" of finite modes *á la* Leibniz.[6] I want to turn our attention, however, to Tschirnhaus since this exchange, interestingly, brings out considerations that go beyond Spinoza.

7.2 The Correspondence with Tschirnhaus

Tschirnhaus raises the following concern in a letter to Spinoza: "First, I find it very difficult to understand how the existence of bodies having motion and figure can be demonstrated *a priori*, since there is nothing of this kind to be found in Extension, taken in the absolute sense" (Ep. 80). In a subsequent letter Tschirnhaus gives us a bit more of a description of what he is expecting Spinoza's geometric method to provide: "In mathematics I have always observed that from any thing considered in itself—that is, from the definition of any thing—we are able to deduce at least one property; but if we wish to deduce more properties, we have to relate the thing defined to other things…" (Ep. 82). What he is demanding then is to begin with something like a definition of Extension, or the adequate idea of infinite extension, and from that, in a series of deductive steps, arrive at ideas of particular bodies with particular figures and motion. However, since on Spinoza's picture there are no other things external to Extension – it is the most general term – Tschirnhaus can't see how Spinoza will be able to deduce a plurality of properties or determinate

---

Hegel's criticism see Samuel Newlands, "Hegel's Idealist Reading of Spinoza," *Philosophy Compass* 6, no. 2 (2011), Yitzhak Melamed, "Acosmism or Weak Individuals? Hegel, Spinoza, and the Reality of the Finite," *Journal of the History of Philosophy* 48, no. 1 (2010) and José María Sánchez de Leon and Noa Shein, "The Coincidence of the Finite and the Infinite in Spinoza and Hegel," *Idealistic Studies* (forthcoming).

[4] This is a claim that is pervasive in and central to Della Rocca's interpretation of Spinoza. See for example Michael Della Rocca, "Spinoza and the Metaphysics of Scepticism," *Mind* 116, no. 464 (2007): 852–54. and *Spinoza*, Routledge Philosophers (London: Routledge, 2008), 6–8.

[5] Curley and Walski, "Necessitarianism Reconsidered," 251.

[6] Ibid., 257.

modes having particular figures and motion from Extension. Recalling that a figure is a bounded piece of Extension and that motion is always relative to other bodies (given that there is no distinction for Spinoza between space and matter), it seems that in principle figure and motion are not in infinite extension, which has nothing external to it to bind it or to move in relation to. Infinite Extension is utterly homogenous, i.e. simple.[7]

Spinoza interestingly first takes Tschirnhaus's question to be tantamount to an objection which can be leveled against Descartes, namely that if Extension is taken to be breadth, length, and depth (extending indefinitely with no vacuum or atoms), it is unclear how Extension can give rise to finite bodies with determinate figure and quantity of motion. In his reply, Spinoza says that, indeed, this is a serious problem for Descartes, but since his own conception of Extension is not identical to that of Descartes, the objection misses its mark. Tschirnhaus, however, presses Spinoza further on this point in his reply: "I should like you to do me the kindness of showing how, from Extension as conceived in your philosophy [i.e. not as an inert mass], the variety of things can be demonstrated *a priori*...." (Ep. 82). Tschirnhaus then is the first to give a precise formulation of the expectation that it be possible to carry out a top-down deduction in Spinoza's system.[8]

Setting to one side Spinoza's actual response for now, one very interesting point that emerges from this exchange is that the problem of the epistemic top-down deduction isn't unique to Spinoza as one might have initially thought given Spinoza's monism and his geometrical method of exposition.[9] Spinoza thinks that Descartes is, in fact, vulnerable to the objection Tschirnhaus raises. This worry highlights something intriguing about Descartes, and, as we shall see, Spinoza as well, although he didn't seem to recognize it. There is a strong sense in which, for Descartes, ontology is supposed to be paralleled by epistemology—we have innate ideas of what there is. If our epistemic trajectory were to parallel exactly the ontological order, we would go from a clear and distinct perception of God to clear and distinct perceptions of thinking substance and corporeal substance (or Thought and Extension), and subsequently to ideas of modes of each of the created substances. That is, we would go top-down, tracking what follows from what, or what depends on what. Yet, this decidedly isn't the way the story goes in Descartes. In the Second Set of Objections to the Meditations compiled by Mersenne, the objector's final request is that:

> ...it would be worthwhile if you set out the entire argument in geometrical fashion, starting from a number of definitions, postulates, and axioms. You are highly experienced in

[7] Note that, for Newton, space is homogenous, infinite etc., and this is precisely everything bodies are not, which in turn he takes to be an argument against Descartes's identification of matter and space. Isaac Newton, *Philosophical Writings*, trans. Andrew Janiak (Cambridge: Cambridge University Press, 2004), 33.

[8] On the effect and involvement of Leibniz in this exchange see Mark Kulstad, "Leibniz, Spinoza and Tschirnhaus: Metaphysics à trois, 1675–1676," in *Spinoza: Metaphysical Themes*, ed. Olli Koistinen and J. I. Biro (New York: Oxford University Press, 2002), 221–40.

[9] I return to discuss Spinoza's actual response at the end of the paper.

> employing this method, and it would enable you to fill the mind of each reader so that he could see everything as it were at a single glance, and be permeated with awareness of the divine power (CSM, II, 92; AT, VII, 128).

Descartes begrudgingly, for reasons we will consider in a moment, offers the beginning of such an exposition. The first proposition of this geometric-style exposition is: "The existence of God can be known merely by considering his nature" (CSM, II, 117; AT, AII, 166). That is, when pressured to provide something like a top-down deduction, Descartes does indeed begin with knowledge of God's existence, and subsequently that God created the heavens and the earth, and can bring about anything that we can clearly and distinctly perceive. Only once this is established does Descartes proceed to prove the real distinction between mind and body. *Prima facie* it seems that proceeding top-down in this fashion is *exactly* what Spinoza insists upon doing in the *Ethics.* Descartes, however, is adamant that this way of proceeding, namely via what he calls synthesis, is ill-suited for doing metaphysics since: "In metaphysics by contrast [to geometry] there is nothing which causes so much effort as making our perception of the primary notions clear and distinct" (CSM, II, 111; AT, VII, 157). Descartes characterizes a synthesis as follows:

> It demonstrates the conclusion clearly and employs a long series of definitions, postulates, axioms, theorems and problems, so that if anyone denies one of the conclusions it can be shown at once that it is contained in what has gone before, and hence the reader, however argumentative or stubborn he may be, is compelled to give assent (CSM, II, 111; AT, VII, 156).

The synthetic method then, has as a prerequisite that we have clear and distinct perceptions of the primary notions from which to proceed with the deduction. Yet, Descartes takes it for granted, and furthermore expects the objectors to concede, that at the outset of our metaphysical investigation, so to speak, we are highly confused precisely about these primary notions. In other words, Descartes is adamant that the first step that must be taken in a metaphysical investigation is doing away with confusion.[10]

Descartes's recognition of the role of confusion in our epistemic journey is crucial for understanding why, *pace* Spinoza, Tschirnhaus's objection misses its mark against both of them. By looking at both Descartes and Spinoza I will offer reasons for holding that a top-down deduction requires a previous bottom-up epistemic trajectory. With these reasons in mind, I will then highlight the importance of acknowledging the role and nature of confusion in Spinoza. Specifically, I will show, first, that confusion is the unavoidable beginning of our epistemic trajectory. More importantly I will articulate the metaphysical underpinnings of confusion, namely, the existence of a plurality of finite modes. This in turn will be an important step in showing that the demand for an absolute top-down deduction is unwarranted, and

[10] Nelson has emphasized the nature of confusion for rationalism more generally in Alan Nelson, "The Rationalist Impulse," in *A Companion to Rationalism*, ed. Alan Nelson (Maldens: Blackwell Pub., 2005), 4–6, 8–9.

the sense in which "conceiving through" is epistemically prior to "following from" for Spinoza.

7.3 Descartes, Extension and the Impossibility of Particular Bodies

Let us begin then by considering why Spinoza thinks Tschirnhaus's objection holds for Descartes. To see this, we should recall that Descartes characterizes Extension, the principal attribute of corporeal substance, as breadth, length and depth (CSM, I, 224; AT, VIII, 42). To clearly and distinctly perceive breadth, length and depth is to clearly and distinctly perceive the essence of corporeal substance. Extension for Descartes has no limits; as he says, it is indefinite and admits of neither vacuum nor atoms (CSM, I, 229-32; AT, VIII, 49–52). It seems that Descartes's extension in and of itself has no internal variation and moreover seems to be inert or deprived of any essential dynamism—it is "merely" length, breadth and depth.[11] In *Principles* II, 25, Descartes, however, claims that a particular body is what is transferred from one vicinity of contiguous bodies, which are regarded as being at rest, to another. What gives rise to particular bodies in Descartes's plenum then is motion. Yet motion, given Descartes's description of Extension, does not seem to be, strictly speaking, part of the essence of matter. Simply put, how could it be possible to deduce motion from simply inspecting a clear and distinct idea of what is essentially inert? Spinoza's first response to Tschirnhaus addresses just this problem: "from Extension as conceived by Descartes, to wit an inert mass, it is not only difficult, as you say, but quite impossible to demonstrate the existence of bodies. For matter at rest, as far as in it lies, will continue to be at rest and will not be set in motion except by a more powerful external cause" (Ep. 81). Spinoza indeed points to the apparent lack of dynamism inherent in Descartes's characterization of the essence of matter as part of the problem. The second part of his comment can be construed as an extra-systematic objection. Spinoza might concede that Descartes, as he seems to do, can allow for God to set matter in motion, and in that way make particular bodies possible. There are, however, two things to note if this is in fact the case. First, this would show that there can't be a deduction top-down from Extension to bodies. At best there would have to be a deduction that would go from God to created substance, and then through God again to modes of substance. That is, God would be

[11] Although this is the standard way among interpreters to think of Descartes's *res extensa*, it is not entirely clear that it is the most favorable way to think about it. A case can be made that extension is regarded as lacking dynamism only when one is considering it while actively excluding the fact that it exists, as the meditator does in Meditation Five. However, when conceiving *res extensa* as an existing substance, it is dynamic. That is, the lack of dynamism is a result not of the essence of extension, but rather of a particular way of regarding it—which involves actively excluding something crucial about it, namely, its existence. This line of interpretation is currently being developed by Alan Nelson and myself. It is clear, however, that Tschirnhaus and Spinoza took Descartes to be committed to an "inert mass".

not only the cause of created substances but also of their modes. This need not imply, of course, two creative acts for God, but if we are looking for deductive steps these two would be separate. Second, Spinoza for metaphysical reasons doesn't think that there can be anything external to Extension; so if Extension were inert as Descartes seems to hold, and there is nothing external to Extension as Spinoza holds, there would be nothing in principle that could set Extension in motion and give rise to bodies. The extra-systematic point is important for understanding Spinoza's reply, but for our purposes it is crucial that the tension arises internally for Descartes. The seemingly unbridgeable gap between Cartesian Extension and Cartesian particular bodies has pushed some, such as Thomas Lennon, to conclude that Cartesian modes of Extension are phenomenal.[12] It is noteworthy that a parallel problem does not arise for modes of Thought. Descartes holds that we have immediate access to the existence of modes of thought; the performance of the *cogito* being the primary example. We cannot doubt there are modes of thought, because an attempt to do so would just be another instance of a mode of thought. Extension presents no such self-evident, indubitable analogue.[13]

It seems then that given Descartes's characterization of the essence of matter, the top-down trajectory is blocked. However, recalling Descartes's insistence that a synthesis requires as a first step a clear and distinct perception of the primary notions, the relevant question to ask regarding the proposed deduction of particular bodies is how do we, in fact, arrive at a clear and distinct perception of Extension (or any other attribute of a created substance for that matter)? We get examples of possible "cognitive routes," as Alan Nelson dubs them, from a particular body to a clear and distinct perception of Extension in the Second Meditation—with the wax example as well as in *Principles*, II, 11—with the inspection of the stone.[14] Although Descartes's ultimate goal in each text is somewhat different, several important considerations emerge. In both cases, Descartes wants the reader to arrive at a clear and distinct perception of Extension. Although his claim that Extension is known through an innate idea might be taken to suggest, *prima facie*, that perceiving it would be the easiest thing for us, these two texts inform us otherwise. Descartes thinks we don't initially have a clear and distinct perception of Extension but are at the outset, rather, highly confused. Our initial confusion is a given; it need not be proven. It is crucial for Descartes that even in a confused idea there is something true—it is not utter falsehood. Descartes's basic strategy is to begin with the confused idea and distill what is true in it. This separating of what belongs to the idea from what is (con)fused with it, is precisely what is involved in gaining a distinct perception or rather in making the idea distinct.[15] It is worth noting Descartes's characterization of clear and distinct perceptions: "I call a perception 'clear' when

[12] Thomas Lennon, "The Eleatic Descartes," *Journal of the History of Philosophy* 45, no. 1 (2007).

[13] A case can be made though that sensations provide us with modes of the union.

[14] Alan Nelson, "Conceptual Distinctions and the Concept of Substance in Descartes," *Protosociology: An International Journal of Interdisciplinary Research* 30 (2013): 197.

[15] For a full elaboration of this point see Alan Nelson, "Descartes's Ontology of Thought," *Topoi: An International Review of Philosophy* 16, no. 2 (1997).

it is present and accessible to the attentive mind—just as we say that we see something clearly when it is present to the eye's gaze and stimulates it with a sufficient degree of strength and accessibility. I call a perception 'distinct' if, as well as being clear, it is so sharply separated from all other perceptions that it contains within itself only what is clear." (CSM, I, 208; AT, VIII, 22). Clear and distinct ideas are contrasted with obscure and confused ideas. Given Descartes's definition of what a distinct idea is as being separated from other perceptions, we can note that confusion is the opposite – namely an agglomeration of ideas.[16]

In the wax example, we begin with the confused idea that the wax is known via its sensible properties. Once, however, we grasp that the wax can lose all of its particular sensible properties and remain itself we recognize that none of the sensible properties are essential to it. Starting with our confused idea of the wax, we note that in addition to what there is to be known in that idea (the wax) there are other ideas fused with it that are separable from it, namely the ideas of the sensory properties: "Let us concentrate, take away everything which does not belong to the wax, and see what is left: merely something extended, flexible and changeable" (CSM, II, 20; AT, VII, 31). Descartes then explains why the flexibility and changeability cannot be known via the imagination. In short, what has been distilled from the initial confused idea (which in fact turned out to be an amalgam of ideas) is the idea of extension. *Principles,* II, 11 presents similar considerations vis à vis an initially confused idea of a stone. Descartes asks us to attend to the idea of the stone and "leave out everything we know to be non-essential to the nature of body" (CSM, I, 227; AT, VIII, 46). He goes on to say: "after all this, we will see that nothing remains in the idea of the stone except that it is something extended in length, breadth and depth." [ibid].

I shall call this process of separating the true content from confusion a bottom-up trajectory: we begin with a confused idea of a particular body and by a process of making the idea ever more distinct ascend to a clear and distinct idea of Extension.[17]

[16]There is of course a wealth of literature on how to understand clarity and distinctness in Descartes. My intention here isn't to provide such an account. For my purposes it suffices to note that distinctness has to do with a separating of ideas, while confusion entails an agglomeration of them. For literature on clarity and distinctness, see for example, Noa Naaman-Zauderer, *Descartes' Deontological Turn* (New York: Cambridge University Press, 2010), 12–17; Kurt Smith, *Matter Matters: Metaphysics and Methodology in the Early Modern Period* (New York: Oxford University Press, 2010), 83–112 and "A General Theory of Cartesian Clarity and Distinctness Based on the Theory of Enumeration in the Rules," *Dialogue* 40, no. 2 (2001); E. M. Curley, "Analysis in the Meditations: The Quest for Clear and Distinct Ideas," in *Essays on Descartes' Meditations*, ed. A. O. Rorty (Berkley: University of California Press, 1986); Lilli Alanen, "Sensory Ideas,Objective Reality, and Material Falsity," in *Reason, Will, and Sensation: Studies in Descartes' Metaphysics*, ed. John Cottingham (Oxford University Press, 1994).

[17]There might be a case to be made here that the *ordo cognoscendi* for Descartes is what he characterizes as "analysis". Nelson and Rogers have recently argued that for Descartes a synthesis is an inverted analysis, and that it requires a previous analysis; see Brian Rogers and Alan Nelson, "Descartes' Logic and the Paradox of Deduction," *The Battle of the Gods and Giants Redux: Papers Presented to Thomas M. Lennon*, ed. Patricia Easton and Kurt Smith (Brill, 2015) 106–136 .While I agree with their interpretation, I don't think what I say regarding the central role of confu-

Once we ascend to a clear and distinct perception of Extension, we can look down, so to speak, and recognize what was true about our initial perception, namely, we can come to recognize correctly the wax and the stone as modes of Extension. We could think of these two directions as corresponding to the order of discovery/*ordo cognoscendi* versus the order of things/*ordo essendi*, sometimes also thought of as a deduction, i.e., a tracking of causal or dependence relations. This two-stage process is prefigured in *Rule Five* of the *Regulae:* "...We shall be following this method exactly if we first reduce complicated and obscure propositions step by step to simpler ones, and then, starting with the intuitions of the simplest ones of all, try to ascend through the same steps to a knowledge of all the rest" (CSM, I, p. 20, AT, X, 379). Although Descartes seems to insist that we must go first bottom-up, from confused to clear and distinct ideas, and then proceed back down from simple clear and distinct ideas, it is crucial to note that the top-down trajectory will not recover our confused idea *qua* confused. That is, the two trajectories are not perfectly symmetrical. For instance, if we began with an idea of a particular body, a stone, say, that included the notion that the stone was independent from other bodies, the idea of its independence, which is confused, will not be recovered. The process of becoming unconfused about our initially confused idea of the stone will make manifest that the stone is not independent but rather only a mode of Extension and that it depends for its delimitation on surrounding contiguous bodies.[18] Although the initial confused idea of the stone conflated modes with substance, the top-down trajectory would be a recognition that the stone is a mode of Extension, which would entail, partially, a recognition that its delimitation depends on the immediately contiguous bodies, and that these depend on their surrounding contiguous bodies, *ad indefinitum*, and that they all, as a totality, follow from Extension. To put this yet another way, the top-down trajectory would put together the idea of Extension with the idea of a mode of extension and the ideas of its surrounding bodies. So, although in both trajectories the initial and ultimate step is an amalgam of ideas, the great difference is that at the outset of a bottom-up trajectory we don't recognize the

sion in what follows hinges upon this. There is, of course, a wealth of views regarding what analysis is exactly for Descartes, but I don't think anything hinges on the identification of analysis and the *ordro cognoscendi* as I've characterized it for the point I want to make. For literature on synthesis and analysis see e.g. Smith, *Matter Matters: Metaphysics and Methodology in the Early Modern Period*; Stephen Gaukroger, *Cartesian Logic* (Oxford: Oxford University Press, 1989); Curley, "Analysis in the Meditations: The Quest for Clear and Distinct Ideas."

[18] There is a debate in the literature as to the number of corporeal substances to which Descartes is committed. In some places he seems to hold that there is a multiplicity of finite extended substances, e.g., a stone (AT, VII, 44), and other places he seems to hold that there is only one extended substance (*Synopsis to the Meditation* CSM II, 10, AT VII, 14). For different views on the issue see Paul Hoffman, "The Unity of Descartes's Man," *The Philosophical Review* 95, no. 3 (1986); Eric Palmer, "Descartes on Nothing in Particular," in *New Essays on the Rationalists*, ed. Rocco J. Gennaro and Charles Huenemann (New York: Oxford University Press, 1999), 26–29; Alice Sowaal, "Cartesian Bodies," *Canadian Journal of Philosophy* 34, no. 2 (2004); Thomas Lennon, "The Eleatic Descartes." *Journal of the History of Philosophy* 45, no. 1 (2007); and Alan Nelson and Kurt Smith, "The Divisibility of Cartesian Extension," in *Oxford Studies in Early Modern Philosophy*, ed. Daniel Garber and Steven Nadler (New York: Oxford University Press, 2010).

amalgam of ideas as the amalgam of ideas it is. Cartesian confusion amounts to this unrecognized amalgamation.[19] Although the terminus of a top-down trajectory is also an amalgam of ideas, unlike at the origin of a bottom-up trajectory, we identify this amalgam correctly as the precise amalgam of ideas it is.

Although the origin and terminus of the epistemological trajectories are not identical, it is important to keep in mind that there is something true in the initial step. Furthermore, the *ordo essendi* must ultimately be the ontological underpinning of the *ordo cognoscendi*. That is, given that there is confusion, this confusion must be explainable or analyzable in terms of the espoused ontology, and ontology of thought, barrowing Nelson's nomenclature again, i.e., the mechanics of ideas.[20] Given that the ontology is what it is, and that we have ideas of what there is, the question is how confusion is at all possible—this is analogous to Descartes's guiding question in the Fourth Meditation. Given that I have a clear and distinct idea of God, and that I perceive him not to be a deceiver, how it is possible I ever err? What I want to highlight here is one of the necessary conditions for confusion, while not providing a full account of its mechanics. What is essential to note about this process is that arriving at a clear and distinct perception of Extension requires a previous latching on to (albeit in a very confused way) a finite mode of extension. This does not entail that substances depend on modes—that would be to confuse the *ordo cognoscendi* with the *ordo essendi*. However, a clear and distinct perception of Extension presupposes, in the *ordo cognoscendi*, some grasp of modes of extension. In other words, what makes a confused perception of the wax or the stone possible is that there is a diversity of extended things. This might be easier to see negatively: if there were no such diversity, then we would immediately clearly and distinctly perceive extension. A clear and distinct perception of Extension, given the way we arrive at a clear and distinct perception of Extension, carries with it implicitly the knowledge that it has finite modes.

Let us return to Tschirnhaus's objection. Descartes can answer by noting that the demand for a purely top-down deduction erroneously disregards something fundamental about our knowledge, namely that we begin with confusion and as a first step must deliver ourselves from that state. Since what makes our initial confusion possible is the existence of a plurality of modes, in a deep sense the plurality is given, albeit unrecognized. Therefore, a demand for an abstractly top-down deduction is unwarranted or artificial. Without a plurality of modes, there would be no amalgamation at all, and hence no confusion. Crucially, we need not prove or deduce that we begin with confusion: we just do begin there.

---

[19] Taking confusion to be a concatenation of clear and distinct ideas is articulated very clearly in Nelson, "Descartes's Ontology of Thought."

[20] Ibid.

7.4 Back to Spinoza

How can Spinoza forestall Tschirnhaus's objection? One strategy would be to show that Spinoza's system enables him to give a response similar to the one I offered on Descartes's behalf. The differences between the Cartesian and Spinozisitic metaphysics and epistemologies notwithstanding, there are important structural similarities. Most importantly, Spinoza concurs with Descartes that we begin our epistemic trajectory in a state of confusion and that attaining adequate knowledge is a process of doing away with said confusion.[21]

It is vital in this context to appreciate why confusion is inevitable for Spinoza such that our epistemic journey begins with delivering ourselves from this state. My intent here is not to give a full account of Spinoza's epistemology or adequate knowledge, but only to point to several key commitments which reveal the metaphysical underpinnings of confusion, which in turn explain why we find ourselves at the outset of our epistemic journey in a state of deep-seated confusion. We can begin by noting, as is well-known, that the human mind for Spinoza is a finite mode of the infinite substance conceived under the attribute of thought. The critical features for our purposes are first, that insofar as the mind is a mode, it is dependent on something else and must be conceived through it; i.e., it is not self-caused and cannot be conceived through itself alone (1D5). Second, insofar as it is finite it is determined or limited by other things of the same kind (1D2), that is, by other finite modes of thought. I have argued elsewhere that the requirement that finite modes be limited by other finite modes together with 1P28 jointly imply that any finite mode is ultimately determined by the totality of finite modes, or, that it necessarily is part of a network of finite modes and that its existence insofar as it is finite is constituted by this determination.[22]

Next we note that the mind being an idea in the infinite intellect must take something as object, and Spinoza provocatively in 2P13 claims that what the mind takes as object is its body. One curious implication of this proposition is that, strictly speaking, the mind has ideas only of its body. However, when we recall that finite modes are determined by other finite modes, this implies that our body is determined by other bodies – proximally by the immediately surrounding bodies, but mediately by the totality of bodies. That is, our body is affected, determined, acted on, by the totality of bodies (both mediate and immediate). Since the body is affected, the mind has ideas of these affections (implication of 2P13 and 2P19). This can be thought of quite literally; the surface determination of a given body is the way in which it is affected by other bodies. The mind, being the idea of the body, has ideas of these affections. One of the cornerstone axioms of the entire work is 1A4, which states: "The knowledge of the effect depends on and involves the

[21] That we are highly confused is established in 2P17-2P31, the rest of Part Two is dedicated to showing how we are to overcome this initial confused state.

[22] Shein, Noa. "Causation and Determinate Existence of Finite Modes in Spinoza." *Archiv für Geschichte der Philosophie* 97, no. 3 (2015), 334–57, and "Not Wholly Finite: The Dual Aspect of Finite Modes in Spinoza." *Philosophia* 46 (2018).

knowledge of its cause." This implies, as Spinoza states explicitly in 2P16, that the idea of an affection of the human body, involves both the idea of the body as well the affecting body. If we think of the surface determination as our paradigmatic example it quite clear why this must be the case. The surface determination is due both to the affected as well as the affecting body.

The ideas we have of the affections of our body, however, are highly confused for several reasons. First, since the mind strictly speaking only takes its body as object, it has an inadequate idea only of the manner in which the body is affected and not what precisely caused the affection. Hence Spinoza says in 2P17 that the body will perceive the affecting body as present until something else comes along to change the affection in the body.[23] At the same time, to know an affection insofar as it is an affection necessarily implies knowing its cause. If we consider what the cause of an affection is, it would be another body, but that body itself would be affected by another body, and so on. That is, since ultimately the cause of an affection insofar as it is a finite thing implicates the full 1P28 network, knowledge of an affection would likewise imply having a grasp on the 1P28 network. This, however, is impossible. Insofar as we are finite, we cannot apprehend the 1P28 network in all its particularity. Or to use Cartesian terminology, we can't have a clear and distinct perception of the 1P28 network. It is precisely because we cannot apprehend the full 1P28 network clearly and distinctly that we perceive the external objects as existing even when they are no longer there.

It seems that finite minds are in a terrible predicament as it seems they cannot free themselves of what we might want to call the circumstance of finitude. That is, it seems that if we take Spinoza at his word regarding 1P28, 2P13 and 2P17, it looks as though we are doomed to utter confusion. However, like Descartes, Spinoza holds that confusion is not falsehood. There is something true in confused ideas (2P35). Note that the structure of Part Two is such as first to explain how, why, and the degree to which we are confused—and only then explain how imaginative ideas may be rationally emended. Given that we are finite minds, and take our bodies as objects, and our bodies are affected by the infinite totality of other bodies, confusion is the unavoidable starting point for cognition.

The fact that we find ourselves confused at the outset, points to the artificiality of expecting to find a direct top-down account. Since we arrive at adequate ideas by emending inadequate ideas, the top-down trajectory cannot be divorced from how we arrive at the top, namely, attaining an adequate idea of an attribute. The way to do that, a *grosso modo*, as Spinoza elaborates in the second half of Part Two, is to recognize those things that are equal in the part and in the whole (2P38). If the cause of confusion was that we needed to somehow get a grip on the infinite particulars of the 1P28 network, 2P38 establishes that knowing *all* the finite modes in their particularity is not necessary. The common notions amount to what in our nature we share with other things. The shared nature is precisely what is both in the part and

[23] In this respect I take Spinoza to by and large be adopting Hobbes's position as to how images are formed. Cf., Thomas Hobbes, *Leviathan: With Selected Variants from the Latin Edition of 1668*, edited by Edwin M. Curley (Indianapolis: Hackett Publishing Company, 1994), 6.

in the whole. The individuation of the modes becomes irrelevant, since the focus is on what is shared. Ultimately, the nature or essence that is shared is that of an attribute, either Thought or Extension. This process isn't entirely dissimilar to what occurs in the wax example—namely noting the shared nature. More broadly though we can say that while for Descartes separation of what is essential from what is not plays a vital role in the cognitive ascent, for Spinoza it is the recognition of a shared nature that pulls us upwards. That is, the mechanics of what gets us through the ascent are different but the goal is the same.[24]

Another way to describe this trajectory is as tracking what the dependent depends on—or to use Spinoza's terminology, conceiving the dependent through what it depends on, or the effect through its cause. That is, the secret to overcoming confusion is noticing that these ideas that are first available to us, ideas of affections of our bodies, depend on others, that is, that they are modes. Conceiving a mode through its causes will ultimately mean conceiving the mode through the substance. A full analysis of this would go beyond the scope of the paper, but all that is needed for our purposes, is that such an analysis can be given.

Once we arrive at (or strive for) an adequate idea of the (formal essence) of an attribute, we can, as we saw in Descartes, reverse the process. This is precisely what Spinoza describes as intuitive knowledge: "...this kind of knowing proceeds from an adequate idea of the formal essence of certain attributes of God to the adequate knowledge of the essence of things" (2P40Schol). That is once we have an adequate idea of the formal essence of an attribute we can see how other things follow from it. Being able to see what follows from what, modes from attribute, requires first that we become unconfused by conceiving what is dependent through what it depends on.

Some commentators have claimed that intuitive knowledge is of individual modes. This however is misleading since it cannot be made to accommodate Spinoza's commitments regarding the nature of finite modes—being determined by other modes. It is worth noting Spinoza's phrasing in 2P40Schol: "the essence of *things*." He does not say "thing" in the singular, but "things" in the plural. He repeats the plural iteration in Part Five as well. In 5P24 we find: "The more we understand singular *things* the more we understand God" (emphasis added). Similarly in 5P36Schol: "...to show by this example how much the knowledge of singular *things* I have called intuitive..." as well as in 5P38: "The more the Mind understands *things* by the second and third kind of knowledge, the less it is acted on by affects which are evil, and the less it fears death." As in Descartes, the terminus of a top-down trajectory is not the confused starting point of an analysis. For Spinoza, although intuitive knowledge is in a way a reversal of the bottom-up trajectory (or the search for common natures), it does not terminate in the confused idea with which the process began. A paradigmatic kind of confusion is treating what is dependent as independent, or rather treating a mode as if it were a substance, i.e., capable of existing independently from other modes. It would be highly curious if intuitive knowledge were to capture precisely what was confused about the initial

[24] I thank Noa Naaman-Zauderer for prodding me on this issue.

idea—i.e. perceiving it both as something particular, or finite, and as independent. The correct view rather, is that intuitive knowledge is a grasping all at once that all things follow from God.

A striking example of how we come to have an adequate idea of an attribute is given in 2P1 where Spinoza proves that Thought is an attribute of God. The demonstration reads:

> Singular thoughts, or this or that thought, are modes which express God's nature in a certain and determinate way (by 1P25C). Therefore (by 1D5) there belongs to God an attribute whose concept all singular things involve, and through which they are also conceived. Therefore, thought is one of God's infinite attributes, which expresses an eternal and infinite essence of God (see 1D6), or God is a thinking thing, q.e.d.

As in the case of Descartes's wax and the stone, we begin our epistemic trajectory with an idea of a mode. What we then note are two things, first that it is a mode, i.e., expresses God's nature in a certain and determinate way. Second, we notice that all ideas share a nature, or as Spinoza says we notice that all ideas involve thought. That is, we come to recognize not only that it is a mode, but that it is a mode *of thought*. Therefore, given that modes express God's nature, ideas *qua* ideas express God's nature and are conceived through it—because they depend on it, and the nature they express is Thought. Ideas then, must be conceived through Thought. Once we do this, we can come to appreciate how all ideas follow from God. Spinoza importantly thinks showing that Extension is an attribute of God proceeds *exactly* along the same lines.

Yet another point of access to this conclusion is the following; my initial self-awareness of my thought or the sensing of body as affected (2A2 and 2A4) might lead me to think that I am an independent thing, have a free will, etc. However, even this highly confused conception indicates that there is a diversity of things, despite utter confusion about their nature. As in Descartes, the reality of confusion indicates the existence of multiplicity. It is an integral part of our epistemic journey that it is self-evident that there is diversity of thought (we have many ideas) and bodies (we sense our bodies affected by other bodies). In other words, the plurality of things is not something to be proven, it is given immediately, albeit confusedly. This givenness does not imply an *ad hoc* assertion of the existence of finite modes. That I am immediately aware that I think and that I sense my body affected does not entail that either one of these states is conceived through itself, or conversely that I cannot conceive its causes. Quite the contrary, the instant I begin comprehending this circumstance I must conceive myself (as a thinking thing or as an extended thing) through something else. To conceive an affection is precisely to conceive the relation of that which is being affected and that which is doing the affecting. If finite modes were established by brute force, they would be unintelligible: we would not be able to track their causes. A core theme, however, in Part Two of the *Ethics* is to show how modes must and can be conceived through their causes and their attributes. Furthermore, when done fully, not only do I conceive myself through my causes, but I also see the way in which I follow from the infinite substance along with all the other finite modes.

7.5 Two Possible Objections

It may be objected contrary to my suggestion here that Spinoza explicitly states in 2P10Schol that the proper order of philosophizing is not bottom-up but rather top-down:

> The cause of this [i.e., those who believe that a created thing can be conceived without God], I believe, was that they did not observe the [proper] order of philosophizing. For they believed that the divine nature, which they should have contemplated before all else (because it is prior both in knowledge and in nature) is last in the order of knowledge, and that the things which are called objects of the sense are prior to all. That is why, when they contemplated natural things, they thought of nothing less than they did of the divine nature; and when afterwards they directed their minds to contemplating the divine nature, they could think of nothing less than of their first fictions, on which they had built the knowledge of natural things, because these could not assist knowledge of the divine nature.

I think however, a closer examination of this passage, together with considerations presented above, reveals he is making a slightly different point. The error that Spinoza is pointing towards is the error of taking sensory knowledge as simple and clear and building up from these supposedly simple and clear ideas others that are more complex. That is, what he is objecting to here is Empiricism. The first point then is that the issue with sensory knowledge is mistaking it for adequate knowledge. This need not conflict with my earlier point that we find ourselves at the outset of our epistemic trajectory amidst confusion. Second, Spinoza is aiming here for a description of what adequate knowledge looks like—so yes, adequate knowledge of particulars must proceed from an adequate knowledge of God—but this leaves it open how we get to an adequate knowledge of God. If anyone could simply start with it, why do so few have it, as he says at the end of Part Five?

Another objection that might be raised against the above proposal is that it looks nothing like Spinoza's actual reply to Tschirnhaus. In Ep. 83 Spinoza replies thus:

> You ask whether the variety of things can be demonstrated a priori from the concept of extension alone. I believe I have already shown sufficiently clearly that this is impossible, and that therefore Descartes defines matter badly by extension, but that it must necessarily be defined by an attribute which expresses eternal and infinite essence. But perhaps I will treat these matters more clearly with you some other time. But as for what you add that from the definition of any thing, considered in itself, we can *only* deduce one property-perhaps this is correct for very simple things, or beings of reason (under which I include shapes also), but not for real beings. For from the mere fact that I define God to be a being to whose essence existence pertains, I infer many of his properties: that he exists necessarily; that he is unique, immutable, infinite, and so on. In this way, I might bring up many other examples, but for the present I will omit them.

What is crucial to note about this response is that Spinoza is setting to one side the difficulties concerning the supposed emergence of finite bodies, by claiming that it is solved by noting that Extension is an attribute of the infinite substance. Spinoza does not explain how this solves the matter but rather leaves it for a further occasion. Spinoza then shifts the discussion to discuss theoretically how it is possible to deduce many properties of a definition of a thing considering it alone, and with this abandons the issue of finite modes. The fact that he is treating the issue of deducing a single property theoretically is evident from his general formulation of the issue, but also most notably from his example. *Prima facie*, it is extremely odd that he offers as the definition of God something other than his own 1D6, and instead offers as a definition "a being to whose essence existence pertains." However, if one notes that the point Spinoza wants to make is a theoretical one—that one can deduce multiple properties by considering a single definition—then it is easy to see why he chooses the example he does. The example allows him to show Tschirnaus that this is common practice by using a very traditional definition of God (which Tschirnhaus would accept) and pointing out that a multiplicity of the traditional divine attributes are said to follow from it. It is clear that Spinoza thinks Tschirnhaus will easily admit as much.

7.6 Conclusion

The state of confusion that marks the origin of our epistemic trajectory presupposes, then, a plurality of modes. Metaphysically speaking, confusion is possible only in virtue of a plurality of modes. At the same time this presupposition is immediately available to us, albeit confusedly: it is self-evident that I perceive a plurality of ideas, and, second, it is equally self-evident that I sense my body affected by other bodies. The demand then for a purely, abstracted top-down deduction is misplaced precisely because it ignores the necessity of first proceeding bottom-up or, rather, overcoming confusion. Overcoming confusion entails a perceptual shift from amalgamation to order and connection; it involves appreciating that modes must be conceived through other mode and/or through the attribute. Once we are successful in doing this, we can see, without confusion that finite modes follow from the infinite substance top-down.

References

Alanen, L. (1994). Sensory ideas, objective reality, and material falsity. In J. Cottingham (Ed.), *Reason, will, and sensation: Studies in Descartes's metaphysics* (pp. 229–250). Oxford: Oxford University Press.

Curley, E. M. (1986). Analysis in the meditations: The quest for clear and distinct ideas. In A. O. Rorty (Ed.), *Essays on Descartes' meditations* (pp. 153–176). Berkley: University of California Press.

Curley, E. M. (1988). *Behind the geometrical method: A reading of Spinoza's ethics*. Princeton: Princeton University Press.

Curley, E. M., & Walski, G. (1999). Spinoza's necessitarianism reconsidered. In R. J. Gennaro & C. Huenemann (Eds.), *New essays on the rationalists* (pp. 241–264). New York: Oxford University Press.

Della Rocca, M. (2007). Spinoza and the metaphysics of scepticism. *Mind, 116*(464), 851–874.

Della Rocca, M. (2008). *Spinoza. Routledge philosophers*. London: Routledge.

Della Rocca, M. (2013). Striving, oomph, and intelligibility in Spinoza. In M. Van der Schaar (Ed.), *Judgement and the epistemic foundation of logic* (pp. 49–65). Dordrecht: Springer Science & Business Media.
Della Rocca, M. (forthcoming). Steps toward eleaticism in Spinoza's philosophy of action. In N. N. Zauderer & T. Vinci (Eds.), *Freedom and the passions in Spinoza's ethics*. Cambridge: Cambridge University Press.
Gaukroger, S. (1989). *Cartesian logic*. Oxford: Oxford University Press.
Hegel, G. W. F. (2010). *The science of logic*. (G. Giovanni, Trans). Cambridge: Cambridge University Press.
Hobbes, T. (1994). In E. M. Curley (Ed.), *Leviathan: With selected variants from the Latin edition of 1668*. Indianapolis: Hackett Publishing Company.
Hoffman, P. (1986). The unity of Descartes's man. *The Philosophical Review, 95*(3), 339–370.
Hübner, K. (2015). Spinoza on negation, mind-dependence and the reality of the finite. In Y. Melamed (Ed.), *The young Spinoza: A metaphysician in the making*. New York: Oxford University Press.
Kulstad, M. (2002). Leibniz, Spinoza and Tschirnhaus: Metaphysics à trois, 1675–1676. In O. Koistinen & J. I. Biro (Eds.), *Spinoza: Metaphysical themes* (pp. 221–240). New York: Oxford University Press.
Lennon, T. (2007). The eleatic Descartes. *Journal of the History of Philosophy, 45*(1), 29–47.
Lin, M. (2006). Substance, attribute, and mode in Spinoza. *Philosophy Compass, 1*(2), 144–153.
Melamed, Y. (2010). Acosmism or weak individuals? Hegel, Spinoza, and the reality of the finite. *Journal of the History of Philosophy, 48*(1), 77–92.
Melamed, Y. (2012). Why is Spinoza not an eleatic monist (or why diversity exists). In P. Goff (Ed.), *Spinoza on monism* (pp. 206–222). New York: Palgrave Macmilan.
Naaman-Zauderer, N. (2010). *Descartes' deontological turn*. New York: Cambridge University Press.
Nadler, S. (2012). Spinoza's monism and the reality of the finite. In G. Philip (Ed.), *Spinoza on monism, Philosophers in Depth* (pp. 223–243). New York: Palgrave Macmillan.
Nelson, A. (1997). Descartes's ontology of thought. *Topoi: An International Review of Philosophy, 16*(2), 163–178.
Nelson, A. (2005). The rationalist impulse. In A. Nelson (Ed.), *A companion to rationalism* (pp. 1–11). Malden: Blackwell Pub.
Nelson, A. (2013). Conceptual distinctions and the concept of substance in Descartes. *Protosociology: An International Journal of Interdisciplinary Research, 30*, 192–205.
Nelson, A., & Kurt, S. (2010). The divisibility of Cartesian extension. In D. Garber & S. Nadler (Eds.), *Oxford studies in early modern philosophy* (pp. 1–24). New York: Oxford University Press.
Newlands, S. (2010). Another kind of Spinozistic monism. *Noûs, 44*(3), 469–502.
Newlands, S. (2011). Hegel's idealist reading of Spinoza. *Philosophy Compass, 6*(2), 100–108.
Newton, I. (2004). *Philosophical writings* (A. Janiak, Trans.). Cambridge: Cambridge University Press.
Palmer, E. (1999). Descartes on nothing in particular. In R. J. Gennaro & C. Huenemann (Eds.), *New essays on the rationalists* (pp. 26–47). New York: Oxford University Press.
Rogers, B., & Alan, N. (2015). Descartes' logic and the paradox of deduction. In P. Easton & K. Smith (Eds.), *The battle of the gods and giants redux: Papers presented to Thomas M. Lennon*. Leiden: Brill.
Sánchez de Leon, José M., & Shein, N. (forthcoming). The coincidence of the finite and the infinite in Spinoza and Hegel. In *Idealistic studies*.
Shein, N. (2015). Causation and determinate existence of finite modes in Spinoza. *Archiv für Geschichte der Philosophie, 97*(3), 334–357.
Shein, N. (2018). Not wholly finite: The dual aspect of finite modes in Spinoza. *Philosophia, 46*, 433–451.

Smith, K. (2001). A general theory of Cartesian clarity and distinctness based on the theory of enumeration in the rules. *Dialogue, 40*(2), 279–310.

Smith, K. (2010). *Matter matters: Metaphysics and methodology in the early modern period.* New York: Oxford University Press.

Sowaal, A. (June 2004). Cartesian bodies. *Canadian Journal of Philosophy, 34*(2), 217–240.

Spinoza, B. de. (1985). *The collected works of Spinoza* (E. M. Curley, Ed. and Trans.). Princeton: Princeton University Press.

Spinoza, B. de. (1995). *The letters* (S. Shirley, Trans.). Indianapolis: Hackett.

Chapter 8
All the Forms of Matter: Leibniz, Regis and the World's Infinity

Mogens Lærke

Abstract In 1697, the publication of a letter from Leibniz to Bourguet in the *Journal des Sçavants* prompted a vigorous reply from the Cartesien Pierre-Sylvain Regis, leading to a public exchange between the two philosophers. The controversy ended with a contribution by Regis who seemingly got the final word. The exchange mainly focused on Descartes's *Principles of philosophy*, III, art. 47, a text where Descartes held that the world would eventually take all the possible forms it is capable of. While Leibniz maintained that this claim had inescapable Spinozistic consequences, Regis defended Descartes against the accusation. Each position relied on its own distinct conception of the world's infinity. This chapter follows this controversy step by step. I show in particular that Leibniz, although he did not write it, had a forceful reply to Regis's final published objections ready at hand, based on his rehabilitation of the notion of actual infinity.

8.1 Introduction

In this paper, I am interested in a controversy between Gottfried Wilhelm Leibniz and the Cartesian Pierre Sylvain Regis, a self-proclaimed representative of orthodox Cartesianism.[1] The controversy stands centrally in Leibniz's sustained attacks on Cartesian metaphysics from the 1679 onward, and it is one of the most public

[1] I use the following abbreviations: DESCARTES: AT = *Œuvres*, 11 vols., ed. C. Adam et P. Tannery, Paris: Cerf, 1897–1909.; CSM(K) = *The Philosophical Writings of Descartes*, ed. and trans. J. Cottingham, R. Stoothoff, D. Murdoch, A. Kenny, 3 vols., Cambridge: Cambridge University Press 1991. LEIBNIZ: A = *Sämtliche Schriften und Briefe*, Berlin: Akademie-Verlag, 1923-; GP = *Die philosophischen Schriften von G. W. Leibniz*, ed. C. I. Gerhardt, Berlin: Weidmannsche Buchhandlung, 1875–1890. The article includes, in condensed form, analyses also presented in French in M. Lærke, *Les Lumières de Leibniz. Controverses avec Huet, Bayle, Regis et More*, Paris: Classiques Garnier, 2015, p. 285–312.

M. Lærke (✉)
CNRS, IHRIM (UMR 5317), Lyon, France
e-mail: mogens.laerke@ens-lyon.fr

O. Nachtomy, R. Winegar (eds.), *Infinity in Early Modern Philosophy*, The New Synthese Historical Library 76,
https://doi.org/10.1007/978-3-319-94556-9_8

controversies with a Cartesian in which he ever engaged. The controversy concerned the question of the world's forms, whether there are infinitely or indefinitely many of them, and what implications either option has for the conception of providence and the modal status of the existing world.

8.2 First Pass: Leibniz's Critique of Descartes

The controversy took place in 1697 and was prompted by a famous letter from Leibniz to Claude Nicaise:

> Even though I am happy to think that Mr. Descartes was sincere in the profession of his religion, the principles that he has laid down still imply some strange consequences that there has not been paid sufficient attention to. After having turned philosophers away from the search for final causes, or what amounts to the same thing, from the consideration of divine wisdom which in my opinion should be philosophy's greatest goal, he gives us a glimpse of the reason in a place of his principles where, trying to justify having arbitrarily attributed certain shapes to matter, he says that he was in his right to do so, because matter successively assumes all possible forms and that it thus must in the end assume the one he supposed. But if what he says is true, that everything possible must happen, and if there are no fictions, however absurd or indignant, that does not happen at some time at some place in the universe, it follows that there is neither choice nor providence, that what does not happen is impossible, and that what does happen is necessary exactly like Spinoza and Hobbes say it in clearer terms. And one can even say that Spinoza did nothing but cultivate certain seeds in Mr. Descartes's philosophy; in such a way that I think it really is important for religion and piety that this philosophy should be reprimanded through the removal of errors that are mixed up with truth.[2]

Written around mid-February, the letter circulated widely in the intellectual circles of Paris, quickly ending up in the hands of Regis who, clearly upset by the comparison of Descartes with Spinoza, set off a controversy in the *Journal des sçavans* as early as June 1697. The public controversy includes four contributions, published over a period of 5 months. The first contribution is a long excerpt from Leibniz's letter, published by Regis in the June 17th issue of the *Journal*, accompanied by a commentary by the Cartesian, which constitutes the second contribution.[3] The third contribution is a long response by Leibniz, published in two parts in the issues of the *Journal* from August 16th and 26th, 1697.[4] The last contribution is a

---

[2] Leibniz to Nicaise, February 15th, 1697, GP II, p. 562–63.

[3] Regis, "Réflexions sur une lettre de Monsieur Leibniz, écrite à Monsieur l'abbé Nicaise dans laquelle il pretend faire voir que les principes de la filosofie de Monsieur Descartes, renferment des consequences contraires à la religion and à la piété," in *Journal de sçavans*, n° 23, June 17th, 1687, p. 273–76 [GP IV, p. 333–36].

[4] Leibniz, "Reponse aux réflexions qui se trouvent dans le 23 Journal des Sçavans de cette année, touchant les consequences de quelques endroits de la filosofie de Descartes," in *Journal des sçavans*, n° 32, August 19th, p. 381–84, and "Suite de la réponse etc.," in *Journal des sçavans*, n° 33, August 26th, 1697, p. 384–88 [GP IV, p. 336–42]. It was Leibniz himself who preferred to publish the text in two parts, deeming it "too long for a journal" (GP II, p. 342).

second response by Regis to Leibniz, published in the *Journal* on November 18th, 1697.[5]

The controversy turns on the question of how to interpret a certain text, namely the following passage in Descartes's *Principia philosophiae*, part III, article 47:

> In fact it makes very little difference what initial suppositions are made, since all subsequent change must occur in accordance with the laws of nature. And there is scarcely any supposition that does not allow the same effects (albeit more laboriously) to be deduced in accordance with the same laws of nature. For by the operation of these laws matter must successively assume all the forms of which it is capable; and, if we consider these forms in order, we will eventually be able to arrive at the form which characterizes the universe in its present state.[6]

In many respects, this is a very marginal text in the general development of the *Principia*. Descartes is in the process of defending the different astronomical principles put in place in the preceding paragraphs and in particular those among them that are directly related to the physics of vortices. In article 47, Descartes insists on the fact that those principles are empirical; they rely on experience. Nonetheless, even from an *a priori* viewpoint, those principles would be valid even when one "start[s] from the primeval chaos <as described by the poets, i.e. a total confusion in all parts of the universe> and deduce from it, in accordance with the laws of nature, the precise organization now to be found in things."[7] For matter will successively take all the forms it is capable of, and what we observe here and now, and which is in general agreement with experience, is thus that we exist in that exact moment in the life of the world where matter has taken on the form which is in agreement with the physics of vortices and the astronomical principles proposed by Descartes.

It is reasonably safe to assume that, for Descartes, this was probably not a crucial piece of reasoning and, it should be said, it is not a very strong one either. Certainly, it has its importance as a means to counter possible Epicurian objections, providing a kind of bulwark against an affirmation of disorder that would render the establishment of an experimental physics impossible. However, putting any strong emphasis on this argument and drawing general conclusions about the Cartesian philosophy from it would not, I think, be considered fair by Descartes, for the argument is at the periphery and not at the center of his doctrine. And yet this is exactly what Leibniz will do, despite the fact that he, in other unpublished papers, had gleaned similar pernicious consequences from texts standing much more centrally in Descartes's philosophy, most importantly the *Replies to the Sixth Objections*.[8] Attacking the Cartesians via a decidedly marginal text did however present a strategic advantage, since it allowed Leibniz to more plausibly argue that his Cartesian adversary was not intentionally, but inadvertently on a slippery slope toward Spinozism, mitigating the accusation somewhat and leaving the adversary with an appropriate excuse.

---

[5] P. S. Regis, "Reflexions pour servir de replique à une reponse insérée dans le 32 & dans le 33 journal de l'année présente," in *Journal des sçavans*, n° 37, November 18th, 1697, p. 439–442.

[6] Descartes, *Principia philosophiae*, III, § 47, AT VIII, p. 103, trans. CSM I, p. 257–58.

[7] AT VIII, p. 102, trans. CSM I, p. 257.

[8] Leibniz, *Periculosa in Cartesio*, 1683–1684/85 (?), A VI, iv, p. 1478.

8.3 Second Pass: Regis's Two Replies

So how did Regis react to Leibniz's letter to Nicaise? First, he replied with an argument *ad hominem*:

> It has long seemed that Mr. Leibniz wants to establish his reputation on the ruins of that of Mr. Descartes. The fragments he has put in the Journal of France from time to time are great proof of this; and the particular relations that he has established with the enemies of this philosopher, who we find here in great number, remove all doubt of it. But the surprising thing is that during all this time not one disciple of Mr. Descartes has undertaken to defend his master. My bewilderment did however end when I learned that [people] have kept silent because they saw that everything Mr. Leibniz wrote destroyed itself and that his best friends openly published that it was preferable that such a great man should confine himself to mathematics where he excels and not get involved in philosophy where he does not have the same advantage.[9]

Regis's complaint was not entirely unjustified. Leibniz was indeed rather cozy with some of the most prominent anti-Cartesians in France, most importantly Pierre-Daniel Huet, whose 1689 *Censura philosophiae cartesianae* still today counts as one of the most ferocious attacks on Cartesian orthodoxy ever produced.[10] As to what concerns Leibniz's contributions to the "Journal of France," that is to say the *Journal des Sçavants*, he had in particular published an article in April 1693 where he aligned himself completely with Huet's onslaught on Cartesianism.[11] I shall not linger on the question why there had been no substantial reaction to Leibniz already in 1693, but for what follows it is important to note how Regis depicts Leibniz as a mere mathematician whose allegedly inept excursions into philosophy eventually self-destruct.

Next, turning to the more substantial part of his reply, Regis attempted to refute point-by-point Leibniz's critique. After first stressing that Descartes did not ban final causes from moral philosophy but only from natural philosophy, he turns to the objection regarding Descartes's inadvertent denial of unrealized possibilities:

> I do not think that, in order to excuse himself from seemingly having arbitrarily attributed to matter certain figures and certain movements, Mr. Descartes has ever said that he had the right to do this because matter successively takes all possible forms. If Mr. Descartes has said this somewhere, I pray Mr. Leibniz to quote the passage, and yet still accept that one believes that Mr. Descartes never proposed such a thing, for one could with good reason have objected that, indeed, matter will at some point arrive at the form that he supposes, but that it has not yet gotten there, and consequently that he explains present things through a future system, which is absurd. Mr. Descartes has never taught that matter successively

---

[9] Regis, "Réflexions sur une lettre," p. 273 [= GP IV, p. 333].

[10] For an English translation, see P.-D. Huet, *Against Cartesian Philosophy*, trans. T. M. Lennon, New York: Humanity Books, 2003. For a commentary, see T. Lennon, *The Plain Truth. Descartes, Huet, and Scepticism*, Leiden: Brill, 2008.

[11] See Leibniz, "Extrait d'une Lettre de M. de Leibnis à M. l'Abbé Nicaise, sur la philosophie de M. Descartes," in *Journal des sçavans* 15, April 13th, 1693, p. 163–165. The original letter is dated June 5th, 1692 (A II, ii, 532–538 = GP II, 534–535). For a discussion of this letter, including a complete English translation, see M. Lærke, "Leibniz, Huet, and the Critique of the Cartesian Spirit," in *The Leibniz Review* 23 (2003), p. 7–36.

> receives all possible forms; he has only said that it can receive them, as it appears from article 203 and 204 of the fourth part of the principles. But if Mr. Descartes does not say that matter successively takes all possible forms, but only those that it can take them, what happens to all the consequences that Leibniz draws from the false doctrine that he attributes to Mr. Descartes? In that case it is no longer true that, according to Mr. Descartes, everything possible must happen, that there is neither choice nor providence, that what does not happen is impossible, that everything that happens is necessary; it is, finally, not true that Mr. Descartes resembles Hobbes or Spinoza: for they absolutely deny divine providence and human liberty; to the contrary, Mr. Descartes expressly teaches in Art. 4 of the first part that it would be a crime to think that we had never been capable of doing something that God had not already commanded.[12]

Regis concluded, concerning Descartes and the Cartesians, that "if it appears to Mr. Leibniz that there is some agreement of their opinions [with Spinozism], this stems not so much from the thing itself as from the false idea that he has conceived regarding Mr. Descartes's doctrine."[13]

8.4 Third Pass: Leibniz's Two-Part Rejoinder

Leibniz's two-part reply appeared in two issues of the *Journal* in August. It begins by apologizing for the publication of the letter, which he explains was not meant for publication. Leibniz insists that he did not intend to accuse the Cartesians of outright Spinozism and expresses his awareness that Descartes explicitly rejected necessitarianism, affirmed providence and allowed explanations from final causes in moral philosophy. He only wanted, he says, to stress certain principles in Descartes, the consequences of which seemed to him to lead in the direction of Spinoza. Now, Regis had not denied at first that those principles had Spinozist consequences. But he had denied that Descartes held them. The whole question, then, turned on what Descartes had in fact said or had not said. Did he in fact claim that matter *takes* all possible forms at some point? Or did he only say that it *could* take all possible forms? Regis denies the former and affirms the latter. But that is, of course, an objection that it was easy for Leibniz to counter by simply "quoting the passage" where Descartes expressly writes what matter must take "all the forms of which it is capable," that is to say, the article 47 of the third part of the *Principia philosophiae*.[14] Certainly, this rebuttal only works on the condition that one understands the term "capable" in the modal sense of "possible." Now, as Jean-Luc Marion has shown, in Descartes, "capable" is a technical term that sometimes, in French (*capable*), acquires a more active that passive meaning, as an actual power

---

[12] Regis, "Reflexions sur une lettre," p. 274–75 [GP IV, p. 334].

[13] *Ibid.*, p. 275 [GP IV, p. 334].

[14] It is not a question that arises among our interlocutors. For another analysis that insist on the "deformation" to which Leibniz allegedly submits Descartes's position, see M. Fichant, "Postface: *Plus ultra*," in G. W. Leibniz, *De l'horizon de la doctrine humaine*, éd. M. Fichant, Paris, Vrin, 1991, p. 188–89, including note 38.

that cannot be reduced to mere potentiality.[15] This is however not a usage that is possible in Latin, and the term Descartes himself uses is indeed the Latin *capax*.[16] Leibniz is thus right about what Descartes in fact said. And behind the apologetic rhetoric in which he swathes it, Leibniz's argumentative strategy and position is very clear. He confronts the Cartesians with an impossible choice: either to continue to ignore the consequences of their own position or to accept and embrace its dangerous implications. The implication is, obviously, that they ought to abandon the Cartesian premises altogether and embrace another position that does not have these unfortunate consequences, such as Leibniz's own.

In this long reply, Leibniz applies all of his strategic acumen. But, given that it was a debate he had been rehearsing for two decades, this is hardly surprising. Hence, already in 1679, a certain Mr. Rabel, a very minor German Cartesian, had asked him the exact same question after having heard Leibniz speak of Descartes's dissimulated necessitarianism during a dinner at the home of Christian Philipp. Shortly after, Philipp wrote to Leibniz with Rabel's complaint: "[Mr. Rabel] was quite appalled, and desires to know the passages where Mr. Descartes positively states those two propositions that you put forward."[17] And already then, Leibniz knew exactly what to reply: "Here you go: *Principiorum philosophicorum part. 3 articulo 47 his verbis*."[18] In fact, we can go even further back: Without any explicit mention of the passage in Descartes, as early as the 1672/73 *Confessio philosophi*, Leibniz denounces "the heresy that believes that in the infinite course of time to come all possibles will be produced at some time or other, and that there is no story that can be imagined that will not come about in the world at some time or other, at least in some slight measure."[19] In two fragments in the *De summa rerum* from December 1675 and December 1676, respectively, Leibniz expresses similar preoccupations, writing that "one must see if it can be proved that there are essences that lack existence, so that it cannot be said that nothing can be conceived which will not exist at some time in the whole of eternity"[20] and stressing that "a God of the kind in whom the pious believe would not be possible, if the opinion of those who believe that all possibles exist were true."[21] As article 47, Leibniz identified it as "dangerous" as early as a letter to Henri Justel from early 1677.[22] Subsequently, Leibniz

---

[15] See J. L. Marion, "De la divinité a la domination: étude sur la sémantique de *capable/capax* chez Descartes," in *Revue philosophique de Louvain*, 73:18 (1975), p. 263–293.

[16] AT VIII, p. 103.

[17] Philipp to Leibniz, 7 (17) January 1680, A II, i, p. 786.

[18] Leibniz to Philipp, January 1680, A II, i, p. 786.

[19] Leibniz, *Confessio philosophi*, 1672–73, A VI, iii, 128–29, trans. in G. W. Leibniz, *Confessio Philosophi. Papers concerning the problem of Evil, 1671–1678*, ed. and trans. R. C. Sleigh. New Haven: Yale University Press, 2005, p. 58.

[20] Leibniz, *De mente, de universo, de Deo*, décembre 1675, A VI, iii, p. 464, trans. in G. W. Leibniz, *De Summa Rerum. Metaphysical Papers 1675–1676*, trans. G. H. R. Parkinson. New Haven: Yale University Press, 1992, p. 7.

[21] Leibniz, Principium meum est., quicquid existere potest, et aliis compatibile est, id existere, December 12th, 1676, A VI, iii, p. 582, trans. in *De summa rerum*, p. 105.

[22] Leibniz to Fabri, early 1677 (?), A II, i, p. 464.

returned to it regularly and always to propose the same criticism.[23] If Regis had followed Leibniz's publications in Latin and in Germany, he could even have read the objection several years before in a published article, for it figures prominently in the *Notata quaedam G. G. L. circa vitam et doctrinam Cartesii*, a text Leibniz wrote in 1689 and published in 1693 in the learned journal *Historia sapientiae et stultitiae* edited by Christian Thomasius in Halle.[24] This said, before 1697, Leibniz had not alluded to the objection in French publications.

8.5 Fourth Pass: Regis on the Indeterminacy of the World

Finally, as the last published contribution to the controversy, Regis's *Réflexions pour servir de réplique à une réponse insérée dans le 32. & dans le 33. Journal de l'année présente* appeared mid-November 1697.[25] The edition of the controversy contained in volume IV of C. I. Gerhardt's *Philosophische Schriften* from the late nineteenth century, which remains the standard but not quite modern edition, does not include this last published text, presumably because Gerhardt was interested only in Leibniz's position and Leibniz does not reply to this last text by Regis. However, if we are to understand the underlying conceptual dynamic driving the controversy, this omission is extremely unfortunate. The *Réflexions pour servir de réplique* has three parts. First, Regis reiterates his argument *ad hominem*, noting that Leibniz "is so accustomed to put Mr. Descartes down that he blames him even when praising him."[26] There is little to say to this: Regis is right and Leibniz knows it. Next follows a part dedicated to the question of final causes where Regis claims that Leibniz "does not take sufficiently into account that created beings can be seen both in relation to their final cause, and in relation to their efficient cause."[27] Regis appears to think that Leibniz is accusing Descartes of focusing exclusively on efficient causes while recommending complete disregard for final causes, and claims that Leibniz ignores the possibility of accommodating both perspectives. The critique is extraordinary in relation to a philosopher who saw it as one of his principal projects to reconcile a mechanistic physics with a finalistic metaphysics. It testifies to the stunning degree to which Leibniz's overall project could be misconstrued by his contemporaries—this is the destiny of a philosopher who wrote much but published little. Finally, as the third and most important part of his response, Regis returns to the question of how to interpret article 47, beginning by making a half-hearted concession:

---

[23] See for example Leibniz to (?), 1679, A II, i, p. 778–779; Leibniz to Philipp, end of January 1680, A II, i, p. 786. See also A VI, iv, p. 1352, 1466, 1478, 1481–82, 1663, etc.

[24] See Leibniz, *Notata quaedam G. G. L. circa vitam et doctrinam Cartesii*, in C. Thomasius, *Historia sapientiae et stultitiae*, vol. II, April–June 1693, p. 113–22 [= A VI, iv, p. 2057–65].

[25] Regis, "Reflexions pour servir de réplique," p. 439–42.

[26] *Ibid.* p. 439.

[27] *Ibid.* p. 440.

> [...] one does indeed find the very same words that Mr. Leibniz employed in his letter to Mr. Abbot Nicaise; but they are not there in the same meaning. According to Mr. Descartes these words mean that matter successively takes all possible forms without however ever being able to arrive at the last (for, according to him, no progression is infinite.) But, from the fact that matter successively takes all the possible forms without ever arriving at the last, what Mr. Leibniz makes Mr. Descartes conclude does not follow. It does not follow, for example, that the same state of the universe must return precisely after some period; it does not follow that all the infinite possibilities must happen within some finite periodical interval; it does not follow that eternity produces nothing new; it does not, finally, follow that everything one can devise, no matter how despicable, must happen. But if all this does not follow, it follows even less *that everything is necessary, and that there is neither choice nor providence*, as Mr. Leibniz accuses Mr. Descartes of having wanted to say.[28]

Let us consider the details of Regis's reply. The essential part takes place in the parenthesis: according to Descartes, he points out, "no progression is infinite." Instead, what Descartes meant in art. 47 was that the succession of forms was never "able to arrive at the last." In short, the succession of "all" forms Descartes had in mind was an *indefinite* progression of forms. Leibniz, however, says Regis, constructs this indefinite progression as if it was an *infinite* one. The difference is that the indefinite, as opposed to infinity, cannot be defined positively, but only negatively as the absence of finitude.[29] By definition, nothing is excluded from an *infinity* of forms: it envelops a totality of forms *to which no other form can be added*. In an *indefinite* progression of forms, on the contrary, there is *always* something that is excluded since, by definition, *another form can always be added*. This is why, for example, a notion like "the greatest number" implies a contradiction: it is in the nature of natural numbers and their indefinite progression that one can always conceive of greater ones, by squaring them, and comparing the relations between the sequences of natural numbers with the sequence of their squares, for example—one enters here all the paradoxes of infinity originally identified by Galileo.[30] What matters in this context, however, is that the indefinite, at every stage of its progression, is determined by the exclusion of what has not yet been included. The indefinite is indefinite by the very fact that one has not yet reached the end and never will. That is why, Regis argues, on the picture of indefinite succession of forms that he attributes to Descartes, there are always forms that have not yet arrived. And among those forms are always included those deemed "most despicable" which thus are possible forms the realization of which is indefinitely deferred in such a way that they, in fact, never come about. For this reason, Regis concludes, it is not true that Descartes destroys providence and choice, for divine providence and choice are expressed in the fact that the most despicable forms never come about. Providence is implicitly defined as a choice of the better defined as the indefinite deferral of the worst: it is, I recall Regis's own words, because "no progression is infinite" but only

[28] *Ibid.* p. 441.

[29] For more on this classic distinction, see the article by A. Schechtman in this volume.

[30] See also O. Nachtomy, "A Tale of Two Thinkers, One Meeting, and Three Degrees of Infinity: Leibniz and Spinoza (1675–78)," in *British Journal for the History of Philosophy* 19:5 (2011), 939–40.

indefinite and thus "without ever arriving at the last" that "it does not follow [...] that everything one can devise, no matter how despicable, must happen" and that, therefore, it does not follow "that there is neither choice nor providence."

The argument can hardly pass for being orthodox. In natural theology, when indefinite progression is invoked in relation to providence, it is habitually to say the opposite: that providence consists in indefinite progress toward the better but without there being a definite endpoint that would be the best. Indeed, as Karl Löwith has shown in his classic commentary, the dynamic, immanentist conception of providence in terms of indefinite progress which, later, came to its full expression in Hegel, already began to appear in the seventeenth century context of the *Querelle des anciens et des modernes*, where the "moderns" began to tentatively propose indefinite progress toward the better as an alternative to the traditional static notion of providential goodness.[31] Regis, of course, was one of the foremost philosophical representatives of the "moderns" in these debates, notably opposed to the anti-Cartesian defender of the "ancients," Pierre-Daniel Huet.[32] The 1691 edition of his *Cours entier de philosophie* was prefaced by a spirited defense of the superiority of Cartesian philosophy, penned by an anonymous author (in fact: Pierre Coste)[33] who developed an argument very close to the defense of modern progress in the arts and sciences proposed by another major representative of the "moderns," Bernard Le Bovier de Fontenelle in his famous 1688 *Digression sur les anciens et les modernes*. This given, on the deeper level of natural theology, one could then expect Regis to advance similar arguments regarding the indefinite progress of goodness, arguing for the providential betterness of the modern world in relation to the ancient one. But this is very far removed from what Regis in fact seems to argue, which is simply that providence consists in an indefinite progression that will always and at every step exclude the worst. Goodness, or betterness at any rate, is eventually obtained, not through the progressive accumulation of goodness, but through the progressive rarefication of worseness, by the constant and necessary exclusion of the worst possible option, which takes place at every bifurcation, determining each and every divine choice. But in all its unorthodoxy, the argument is not inept. Indeed, one is left with the impression that Regis has scored a point and refuted Leibniz well.

---

[31] See K. Löwith, Meaning in History: The Theological Implications of the Philosophy of History, Chicago: University of Chicago Press, 1957, p. 60–61.

[32] On the disputes between Regis and Huet, see P.-D. Huet, *Censura philosophiae cartesianae*, Paris: Daniel Horthemels, 1689; P.-S. Regis, Pierre-Sylvain, *Reponse au livre qui a pour titre P. Danielis Huetii, Censura philosophiae Cartesianae*, Paris: Jean Cusson, 1691; P.-D. Huet, *Censura philosophiae cartesianae*, 4e éd. *aucta et emendata*, Paris: Joannem Anisson, 1694. For commentary, see T. M. Lennon, *The Plain Truth. Descartes, Huet, and Scepticism*, Leiden: Brill, 2008.

[33] Anon. [in fact: P. Coste], "Discours sur la philosophie, ou l'on voit en abrégé l'histoire de cette science," in P.-S. Regis, *Cours entier de philosophie ou Systeme general selon les principes de M. Descartes*, Amsterdam, Huguetan, 1691b, vol. I, not paginated, p. 1r-22v.

8.6 Fifth Pass (Conjectural): Leibniz on the Infinity of the World

Leibniz did not respond. The question that arises then is this: Was this because he recognized himself defeated? Did he capitulate to Regis's unorthodox but seemingly ingenious conception of providence? Certainly not!

First, note the direction in which the controversy is now moving. In his first response to Leibniz, Regis attempted to ridicule Leibniz by depicting him as a mathematician clumsily contributing to a discipline where he "doesn't have the advantage," i.e. philosophy, suggesting that he should stick to his own area of competence before self-destructing. Well, now the controversy is moving exactly in that direction, toward mathematics and the interpretation of infinity, courtesy of Regis himself. Indeed, from a strategic point of view, and inadvertently no doubt, Regis hands Leibniz an astounding advantage, once again revealing his ignorance of Leibniz's previous preoccupations, in mathematics and in philosophy. For this alleged premise regarding the impossibility of "infinite progression," stated by Regis in a parenthesis as uncontroversial, had been subjected to intense mathematical and philosophical scrutiny by Leibniz who had found it incorrect. In fact, confirming once again Regis's neglect in checking his adversary's published work (without however abstaining from dismissing it!), Leibniz had clearly presented his position in the *Journal des sçavants* in an excerpt from a letter to Foucher published in August 1693:"I am so much in favor of actual infinity that, instead of admitting that nature abhors it, as it is commonly said, I hold that it is affected by it everywhere, in order to better mark the perfections of its author."[34] This bold affirmation rested on reflections going back two decades. Indeed, his rejection of the Cartesian position, which on that point simply followed Aristotelian orthodoxy, and his rehabilitation of "actual infinity" went as far back as to the mid-1670s, to the time of the discovery of the calculus (end 1675) and coincided, more or less, with his reading and commentary on Spinoza's famous "Letter on the Infinite" (around February 1676). In this letter, Leibniz found a rare ally in his fight for rehabilitating actual infinity, a notion rejected as contradictory both by the Aristotelian and the Cartesian tradition, but embraced by the Jewish anti-Aristotelian Chasdaï Crescas against Maimonides and, via Crescas,[35] also adopted by Spinoza.[36]

All this to say that, courtesy of Regis, the whole controversy was now displaced toward a theoretical domain where Leibniz was much more at ease than Regis. And

---

[34] Leibniz, "Reponse de M. de Leibniz à l'extrait de lettre de M. Foucher Chanoine de Dijon, insérée dans le Journal du 16 mars 1693," in *Journal des sçavans*, n° 30, August 3rd, 1693, p. 355–356, here p. 356 [the original letter from around January 1693 can be found in GP I, p. 415–416).

[35] See H. A. Wolfson, *Crescas' Critique of Aristotle*, Cambridge, Mass.: Harvard University Press, 1971.

[36] See M. Lærke, "Leibniz's Cosmological Argument for the Existence of God," in *Archiv für Geschichte der Philosophie* 93 (2011), 59–68. On Spinoza, Crescas and infinity, see also M. Lærke, "Spinoza and the Cosmological Argument According to Letter 12," in *British Journal for the History of Philosophy* 21:1 (2013), p. 57–77.

for Leibniz, as a mathematician, denying, as Regis did, "infinite progression" in a parenthesis amounted to exactly the kind of unquestioned deferral to scholastic orthodoxy that Cartesianism was supposed to be the antidote to. It was an unexamined—and for Leibniz mistaken—adoption of the deeply entrenched Aristotelian principle *infinitum actu non datur.*

So how exactly could Leibniz have responded to Regis's attempt at shielding Descartes by interpreting the notion of "all forms" in terms of an "indefinite progression" of forms? He could have replied in at least three ways. The first relates to the consistency of Regis's reading with his own previous views; the second to its conformity with the authentic Cartesian doctrine of which he claims to be the representative; the third to the tenability of the position as such as argument against necessitarianism and in favor of providence.

First, he could have objected that Regis was philosophically opportunist, contradicting even himself for the occasion. In his grand Cartesian *summa* from 1691, the *Système de philosophie* (also published as *Cours entier de philosophie*), Regis discussed Descartes's vortex theory in some detail, including article 47.[37] And here, he reconstructed Descartes's position as meaning that God has given matter an "infinity of variations":

> [...] God has not divided the extension of the world in such a way that there is any empty space between the parts but [...] all of divisions that he has established between them consists in the movement he has imparted to each in such a way that, from the first moment when they have been moved, some have begun to go toward one side, and others to another; from which, if we add to this that the parts of matter are impenetrable, it appears evident that God could not have continued to move that without there occurring an infinity of variations in their movements, until they finally all agreed to move around different centers.[38]

On this reading, when speaking of "all the forms of which [matter] is capable" in art. 47, Descartes meant that matter would assume an infinite variety of forms. Now, in terminological rigor, this reading is at odds with Regis's own reply to Leibniz since he here in fact speaks of an infinity of forms which he explicitly rejects in the reply. This is however not a very strong objection to Regis. There is little in Descartes to support the view that by using the term "all forms" in art. 47, he broke with an Aristotelian orthodoxy that he clearly follows elsewhere. There is even less evidence that Regis intended to stray from the letter of Descartes's doctrine. Indeed, I think we can safely assume that the term "infinity" here reflects only carelessness in writing, and that Regis's considered view already in the *Système* was to understand "all forms" in terms of an indefinite succession of forms.

Next, from the point of view of the broader Cartesian position, Leibniz could have noted that Regis must have misunderstood Descartes's meaning, because his interpretation of art. 47 does not allow one to reach the conclusion that Descartes draws from it. For, in case "all forms" means indefinitely many forms, nothing prevents maintaining that the actualization of the state of the world that agrees with

[37] P. S. Regis, Cours entier de philosophie ou Systeme general selon les principes de M. Descartes, Amsterdam: Huguetan, 1691b, t. II, ii, chap. 4–12, p. 397–424.

[38] *Ibid.* II, ii, chap. 4, p. 398.

physics of vortices could be deferred indefinitely and thus that that particular world might be the one that never comes to be. And there is no given *a priori* reason why that world should not be the one that would be deferred, no proof that it is in fact not deferred other than observing what is the case, i.e. that the physics of vortices works. Consequently, on Regis's reading, Descartes's presumed *a priori* argument falls apart: for on this reading of Descartes, there is no absolute necessity or *a priori* reason that the matter configured in conformity with the physics of vortices will in fact find itself realized at some point. The alternative interpretation of art. 47 that Regis proposes in response to Leibniz's objection thus deprives Descartes's argument of its anti-Epicurean force, ignoring the explicit argumentative aim set for the article by Descartes himself. This is hardly an appropriate strategy for someone who presents himself as the faithful heir to Descartes; for Regis would only save Descartes by betraying him.

Finally, and most importantly, from Leibniz's own viewpoint, external to Cartesianism, he could give the following response. Let us suppose that Descartes indeed meant "indefinitely many" when speaking of "all forms." Leibniz would then reply that such a world of indefinitely many forms would never come into existence. It would never become real. For in fact, and contrary to what Descartes held, only actual infinity can exist, and all that exists is actually infinite. Indeed, Leibniz could reply, *not only is infinity actual, but there is nothing actual that is not infinite*. The infinity of the world is a necessary condition for its actuality. Or, put another way, infinity is a necessary condition for existence. The indefinite, on the contrary, will never give rise to a real world. Why? Because a world actualizing itself indefinitely would have no *reason* for existing. For, by its very nature, what is indefinite cannot be conceived in terms of an *optimum*, since more can always be added to it. Consequently, *a world of indefinitely many forms cannot be the best*. And for this reason God has no reason to create it.

Let us try to spell out the argument in more detail. Regis argues that, on Descartes's view, the actualization of the world proceeds indefinitely, without however the worst option ever coming about. Divine providence consists in this constant avoidance of the worst. But in that case, Leibniz could reply, Regis also commits himself to holding that, in an important sense, the best is never actualized either, because it is of the nature of the indefinite itself that one can always add to it. Take any indefinite progression of good things, it will never be the best conceivable, for yet another good thing can always be added, and to that, yet another, and so on. If Regis is arguing, as it appears that he does, that the workings of providence in divine creation reduces to the mere fact that the created world never is, at any point, the worst, this also ends up implying that it is never the best either, because on this conception of the world, as an indefinite progression, bestness in any absolute sense is just not conceivable. But in that case, why should God create that world at all? It seems odd to claim that God created and continually recreates a world to his own glory simply because it is not the worst. Certainly this at most provides a reason why God created this world rather than another, worse one, but not a reason why there is a world at all. And, by Leibniz's lights, a reason is required for both. As he writes later in the *Principes de la nature et de la grâce*:

> [...] now we must rise to metaphysics, by making use of that great principle, little used, commonly, that nothing takes place without sufficient reason, that is, that nothing happens without it being possible for someone who knows enough things to give a reason sufficient to determine why it is so and not otherwise. Assuming this principle, the first question we have the right to ask will be, why is there something rather than nothing? For nothing is simpler and easier than something. Furthermore, assuming that things must exist, we must be able to give a reason why they must exist in this way, and not otherwise.[39]

In order for the world to exist, then, there *has* to be infinity: the "infinite progression" that Regis rejects *must* exist in order for the world to exist. And if one persists in rejecting the actual infinity of the world, this amounts to affirming that the world exists without enveloping an *optimum*, and thus the reason of the best. To say, however, that the world thus *exists for no good reason* entails all the consequences for morality and the use of final causes that Leibniz deplores, consequences that Cartesians such as Regis sought as much as Leibniz to avoid. Indeed, in his attempt to avoid the necessitarianism of Spinoza by invoking the indefinite, Regis inadvertently ends up embracing a kind of voluntarism which, from the moral point of view, is equivalent to necessitarianism through a kind of *coincidentia oppositorum*. As Leibniz writes already in the mid-1680s in the *Periculosa in Cartesio*, about Descartes's position in the *Replies to the Sixth Objections*:

> [I]t can be proven by no argument that God is good and truthful, if the very reason of truth and goodness depends on the arbitrary [choice] of God [*Dei arbitrio*]. This is why I am more inclined to think that Descartes's deepest convictions are not too far from Spinoza's opinions.[40]

What is required to counter such a position is a conception according to which forms actualize infinitely, or rather a world where the infinity of forms would provide the very reason of their actuality, but in such a way that this actual infinite does not include *all* possible forms. Only one among infinitely many possible infinite series, successions or progressions, of the world can be actual, namely the best. And this, of course, is exactly the kind of conception that Leibniz, in 1697, had been working on for decades in order to provide an alternative to Descartes and Spinoza. Indeed, 18 years earlier, in 1679, he already had written to Christian Philipp:

> As to what regards Descartes's philosophy that you ask my opinion about, I am wary to say absolutely that it leads to atheism. It is true that there are certain things that appear deeply suspect to me who has considered it carefully. For example these two passages [according to which] one should not consider final causes in physics, and that matter successively takes on all the forms of which it is capable [...]. As for me, I believe [...] that matter does not successively take all the forms possible but only the most perfect ones.[41]

In 1697, then, Leibniz had at his disposition an enormous metaphysical and mathematical apparatus of argumentation in place, allowing him to rehabilitate the notion that Regis rejected in a parenthesis as beyond discussion, that is to say, the

[39] Leibniz, *Principes de la nature et de la grâce*, § 7, GP VI, p. 602, trans. in G. W. Leibniz, *Philosophical Essays*, eds. R. Ariew and D. Garber, Indianapolis: Hackett, 1989, p. 209–10.

[40] Leibniz, *Periculosa in Cartesio*, 1683–1684/85 (?), A VI, iv, p. 1478.

[41] Leibniz to Philipp, December 1679, A II, i, p. 767.

notion of an actual infinite progression. In short, at this point of their controversy, the prospects for Regis were not good and one could expect Leibniz to triumphantly give his unsuspecting adversary the *coup de grâce* with a last devastating lesson on the intricacies of actual infinity.

8.7 Conclusion: Why Did Leibniz Not Reply?

But he didn't. Why? We find a succinct summary of the entire controversy in a letter to Father Verjus, a Jesuit, in December 1697:

> I should say on this occasion to you, reverend father, that, after having written a letter to a friend, who was also a friend of the Bishop of Avranches, and where I proposed that it appeared to me good and necessary to put serious restrictions upon what is taught in Cartesian philosophy, this letter was rendered public in the *Journal des sçavans* in Paris, and that it is refuted there is as if I had attacked the religion of the late Mr. Descartes and the Cartesians. It is true that I had noted in this author some propositions the outcome of which seemed to do harm to providence and to lead to the doctrine of Spinoza. But I had at the same time declared that I did not impute these outcomes neither on the author nor on his followers, since charity obliges me to believe that they have not noticed these dangerous consequences. And since my anonymous antagonist objected to a proposition which I said could be found in Descartes, according to which matter successively receives all possible forms, and from which I inferred that since everything possible would happen there would be neither choice nor providence, I did not omit to point to the place in Descartes, to wit, the 47th article of the 3rd book of the *Principles*, and to justify my remark in the 32nd and 33rd *Journal* from this year. Most recently, there has been a reply in the 37th Journal where, without taking into account my formal declaration I am still accusing the religion of this author; which is unjust. I say these things so that you, reverend father, will recognize my good intentions as well as my moderation.[42]

Apparently, Leibniz worried that a prolonged controversy would only contribute to bring Regis and himself even further from each other, and any insistence on his part be conceived as one long prolonged attack on the religion of Descartes. One can only guess that instead, by keeping silent, he rather preferred to make a display of "good intentions" and "moderation" rather than the superiority of reason.

Over and above the mundane question of who came out victorious from their 1697 controversy, Leibniz's exchanges with Regis is a good example of how a very abstract discussion regarding the nature of infinity could very quickly take turn toward seemingly far removed questions of regarding moral philosophy and natural theology. Hence, in this case, Regis' appeal to the Cartesian notion of indefinite progression and Leibniz's embrace of actual infinity gave rise to sharply differing views on the nature of divine providence.

[42] Leibniz to Verjus, 2 December 1697, A I, xiv, p. 839–40.

Abbreviations

A	*Sämtliche Schriften und Briefe*, Berlin: Akademie-Verlag, 1923– .
AT	*Œuvres de Descartes*, 11 vols., ed. C. Adam et P. Tannery, Paris: Cerf, 1897–1909.
CSM	*The Philosophical Writings of Descartes*, ed. and trans. J. Cottingham, R. Stoothoff, D. Murdoch 2 vols., Cambridge: Cambridge University Press 1984–5.
CSMK	The Philosophical Writings of Descartes, ed. and transl. by J. Cottingham, R. Stoothoff, D. Murdoch, and A. Kenny, vol. 3: *The Correspondence*, 1991.
GP	*Die philosophischen Schriften von G. W. Leibniz*, ed. C. I. Gerhardt, Berlin: Weidmannsche Buchhandlung, 1875–1890.

References

Fichant, M. (1991). Postface: Plus ultra. In M. Fichant (Ed.), *Leibniz, G. W. De l'horizon de la doctrine humaine* (pp. 125–210). Paris: Vrin.

Huet, P. D. (1689). *Censura philosophiae cartesianae*. Paris: Daniel Horthemels.

Huet, P. D. (1694). *Censura philosophiae cartesianae, 4e éd. aucta et emendata*. Paris: Joannem Anisson.

Huet, P. D. (2003). *Against Cartesian philosophy* (T. Lennon, Trans.). New York: Humanity Books.

Lærke, M. (2003). Leibniz, Huet, and the critique of the Cartesian Spirit. *The Leibniz Review, 23*, 7–36.

Lærke, M. (2011). Leibniz's cosmological argument for the existence of God. *Archiv für Geschichte der Philosophie, 93*, 59–68.

Lærke, M. (2013). Spinoza and the cosmological argument according to letter 12. *British Journal for the History of Philosophy, 21*(1), 57–77.

Lærke, M. (2015). *Les Lumières de Leibniz. Controverses avec Huet, Bayle, Regis et More*. Paris: Classiques Garnier.

Leibniz, G. W. (1989). In R. Ariew & D. Garber (Eds.), *Philosophical essays*. Indianapolis: Hackett.

Leibniz, G. W. (1992). *De Summa Rerum. Metaphysical papers 1675–1676* (G. H. R. Parkinson, Trans.). New Haven: Yale University Press.

Leibniz, G. W. (2005). *Confessio Philosophi. Papers concerning the problem of Evil, 1671–1678* (R. C. Sleigh, Ed., and Trans.). New Haven: Yale University Press.

Lennon, T. M. (2008). *The plain truth. Descartes, Huet, and Scepticism*. Leiden: Brill.

Löwith, K. (1957). *Meaning in history: The theological implications of the philosophy of history*. Chicago: University of Chicago Press.

Nachtomy, O. (2011). A tale of two thinkers, one meeting, and three degrees of infinity: Leibniz and Spinoza (1675–78). *British Journal for the History of Philosophy, 19*(5), 939–940.

Regis, P.-S. (1691a). *Reponse au livre qui a pour titre P. Danielis Huetii, Censura philosophiae Cartesianae*. Paris: Jean Cusson.

Regis, P.-S. (1691b). *Cours entier de philosophie ou Systeme general selon les principes de M. Descartes*. Amsterdam: Huguetan.

Wolfson, H. A. (1971). *Crescas' critique of Aristotle*. Cambridge, MA: Harvard University Press.

Chapter 9
Leibniz's Early Encounters with Descartes, Galileo, and Spinoza on Infinity

Ohad Nachtomy

Abstract This chapter seeks to highlight some of the main threads that Leibniz used in developing his views on infinity in his early years in Paris (1672–76). In particular, I will be focusing on Leibniz's encounters with Descartes, Galileo, and Spinoza. Through these encounters, some of the most significant features of Leibniz's view of infinity will begin to emerge. Leibniz's response to Descartes reveals his positive attitude to infinity. He rejects Descartes's view that, since we are finite, we cannot comprehend the infinite and therefore should refrain from studying it. Likewise, Leibniz rejects Descartes's view that the term 'infinite' should be reserved to God alone, as well as Descartes's distinction between the infinite and the indefinite. Leibniz's encounter with Galileo brings out his rejection of infinite number in response to Galileo's paradox. This, in turn, leads him to face another formidable challenge, viz., to defend the claim that an infinite being is possible, while an infinite number is not. Leibniz's encounter with Spinoza, I suggest, highlights the way he is approaching this problem by distinguishing between quantitative and non-quantitative senses of infinity. The strategy of employing different senses of infinity in different contexts will remain central in Leibniz's approach to infinity for the rest of his career.

9.1 Introduction: Leibniz and the Significance of His Encounters

In some of his autobiographical remarks Leibniz describes how, at a very tender age, he taught himself to read Latin through an illustrated edition of Livy's Roman history that he found in his father's library. Once allowed into the library, he went

This research was supported by the Israel Science Foundation (grant 302/16).

O. Nachtomy (✉)
Philosophy Department, Bar-Ilan University, Ramat Gan, Israel
e-mail: ohadnachtomy@mac.com

O. Nachtomy, R. Winegar (eds.), *Infinity in Early Modern Philosophy*,
The New Synthese Historical Library 76,
https://doi.org/10.1007/978-3-319-94556-9_9

through his father's books with great curiosity and pleasure.[1] Leibniz's vast erudition is clearly one of his distinctive traits. At the same time, he was not a typical erudite, a scholar with a gift for accumulating and memorizing what he reads. Leibniz was gifted with a rare ability to quickly adapt and modify what he read. It is no wonder therefore that Leibniz has often been described as an eclectic. But Leibniz was not an ordinary eclectic. Rather, his approach to his many sources marks a definite pattern. Indeed, one of Leibniz's most distinctive intellectual traits – and one of his most remarkable talents – was his ingenious ability to integrate and synthesize views that seem far apart, or even opposed. By reconciling such seemingly conflicting views, Leibniz was producing original views that bear the mark of his synthetic and creative mind.[2]

His conciliatory approach to his sources was further motivated by his conviction that each view contains (at least) a part of the truth. As he tells Remond in 1714:

> I have found that most of the sects are right in a good part of what they propose, but not so much in what they deny... I flatter myself to have penetrated into the harmony of these different realms and to have seen that both sides are right provided that they do not clash with each other... (GP III 606; L 654-5).

Thus, for Leibniz, the challenge was to produce a framework that would harmonize apparently conflicting views.[3] In addition, Leibniz was one of the most prominent citizens of the "Republic of Letters". His lively and voluminous correspondence testifies for the dialogical character of his thinking and the way he engages with other thinkers' work.[4] But precisely because Leibniz read so much and conversed with so many, trying to single out *his ultimate* sources and *the* readings that shaped his views is not only impossible but also a misguided approach. And yet, since Leibniz is engaging with works of others through his conciliatory approach, looking at Leibniz's intellectual encounters is immensely rewarding. In my view, this is the

---

[1] Although his autobiographical account seems rather incredible, it is probably only slightly exaggerated. For a balanced, and well-informed account of Leibniz's intellectual formation, see Antognazza (2009), especially pp. 30–37. While most accounts highlight Leibniz as a miraculous autodidact, Antognazza's account balances this with his formal schooling: "At school he was drilled for hours in textbook fashion on a tiny canon of established authorities in a restrictive elementary curriculum. At home he spent entire days wandering freely in the intellectual terra incognita of this father's library..." (35).

[2] One early example of Leibniz's intellectual attitude is the construction of his early vision in the project for catholic demonstrations and his 1668 *confessio Naturae contra Atheistas*. See next section for more details and footnote 7, in particular.

[3] It is worth noting that, according to Leibniz, harmony itself is seen as unity in diversity. The epistemic sense of harmony – a sense of forming a unified view of various sources – would ultimately track the harmony, which, according to Leibniz, was the very principle of creating the world (and thus the harmony among created things). To produce harmony among diverse views – that is, to achieve order and unity among different views and various disciplines – was clearly one of Leibniz's central aims.

[4] One of the most telling facts illustrating Leibniz's attitude, apart from the sheer volume of his letters, is his composition of the imaginary dialogue with Locke, resulting in the "New Essays on Human Understanding," due to his frustration with Locke's lack of response to his attempts to draw him into dialogue.

best way to grasp the subtlety and complexity of his views, as long as we keep in mind that we shall not exhaust the sources that figure in his work and that his approach to his sources is creative and productive in serving the formation of his own views. It is with this in mind that I propose to study Leibniz's early views on infinity by presenting some of the (many) encounters that shaped his views. In what follows I present Leibniz's early views of infinity as emerging from his engagement with Descartes, Galileo, and Spinoza.

Encounter is a loose enough word for my purposes here. I use it to designate Leibniz's engagement with the texts and views of three major thinkers that play important roles in the formation of his views. I do not attempt here a thorough presentation of Leibniz's exchanges; rather, I seek to highlight and trace some of the main threads that Leibniz used in developing his views on infinity. In particular, Leibniz's early response to Descartes reveals his positive attitude to infinity. He rejects Descartes's view that, since we are finite, we cannot comprehend the infinite and therefore should refrain from studying it. Likewise, Leibniz rejects Descartes's view that the term 'infinite' should be reserved for God alone and seeks a way to use infinity both in mathematics and in metaphysics. Leibniz's encounter with Galileo brings out his rejection of infinite number in response to Galileo's paradox. This, in turn, leads him to face another challenge, viz., to defend the claim that while an infinite number is impossible, an infinite being is possible. Leibniz's encounter with Spinoza, I suggest, highlights his way of approaching this problem by distinguishing between different senses of infinity.

In particular, the most significant distinction Leibniz needs to face this challenge is between a quantitative and a non-quantitative sense of infinity. The quantitative sense applies to magnitudes (and is mainly used in a syncategorematic sense). Leibniz's syncategorematic reading is clearly expressed in the following passage from the *New Essays in Human Understanding*:

> It is perfectly correct to say that there is an infinity of things, i.e., that there are always more than one can specify. But it is easy to demonstrate that there is no infinite number, nor any infinite line or other infinite quantity, if these are taken to be genuine wholes. The Scholastics were taking that view, or should have been doing so, when they allowed a 'syncategorematic' infinite, as they called it, but not a 'categorematic' one. (NE 2.17.1, p. 157)

Richard Arthur has argued convincingly that Leibniz developed this interpretation during his stay in Paris. As he notes in his recent book,

> by the time he had finished laying the groundwork for his calculus in 1676, he already had an interpretation of the mathematical infinite and the infinitely small as *fictions* – as a means of abbreviating statements about an arbitrary large number of things of arbitrary smallness. Just as the infinite is not an actually existing whole made up of finite parts, so infinitesimals are not actually existing parts which can be composed into a finite whole. (Arthur 2014, p. 81)[5]

In what follows, I will suppose the syncategorematic reading and Arthur's interpretation of it. Since much has been written about this distinction, I focus on other

[5] See also Leibniz to Varignon, 2 Feb. 1702, GM IV 93, L 543; Leibniz to Bernoulli, 7 June 1698, GM III 499, L 511.

distinctions, which are of no less importance, here. In particular, the comparison between an infinite being and infinite number is especially helpful to clarify the sense in which infinity applies to the most perfect being. According to Leibniz, the most perfect being does not admit of part/whole relations; rather, it has absolute unity and perfection, which need not and cannot be measured or understood in quantitative terms. Leibniz's encounter with Spinoza reveals an intermediate degree of infinity as well, which he calls "maximum in its kind". But this aspect of Leibniz's view can be only mentioned here. I should stress that the above distinctions are used in addition to (and are consistent with) Leibniz's distinctions between a categorematic infinite, which he rejects; a syncategorematic infinite, which he accepts; and the hypercategorematic infinite, which he applies to God alone.[6] These distinctions are also consistent with Leibniz's other familiar distinction between actual and potential infinity. Leibniz's strategy, employing different senses and different degrees of infinity in different contexts of his philosophy, is at the heart of his approach to infinity. I draw a more comprehensive map of these distinctions elsewhere.[7] Here I can only set up the background and Leibniz's main motivation for distinguishing a quantitative and a non-quantitative sense of infinity.

9.2 Leibniz's Encounter with Descartes

When Leibniz made his first steps into the scholarly world, Descartes's fame was already well established, and his work widely disseminated throughout Europe. It would have been surprising, therefore, if Leibniz, who was keen to read whatever he could of the new philosophers, had not been familiar with Descartes's work. However, as Maria Rosa Antognazza notes in her study of Leibniz's intellectual biography, "although [Leibniz] was obviously familiar with Descartes's philosophy, his knowledge of it up to [1675] had been basically second-hand. During the winter of 1675–6 and the spring of 1676 he plunged into a careful reading of Descartes's *Principia Philosophiae* (Amsterdam 1644), leaving after him a trail of notes."[8]

As his notes testify, one thing that particularly drew Leibniz's attention was Descartes's distinction between the infinite and the indefinite. In Part I, articles 26–27, of his *Principles of Philosophy*, Descartes distinguishes between the infinite and the indefinite, and argues that we should not seek to comprehend the infinite, but should rather consider what we find without limits to be *indefinite* (article 26,

---

[6] In a letter to Des Bosses in 1706 Leibniz writes:

"In addition [to the syncategorematic infinite], there is a hypercategorematic infinite, or potestative infinite, an active power having, as it were, parts eminently but not formally or actually. This infinite is God himself" (DesB 52–53; GP II, 314–315).

[7] In my forthcoming book manuscript Living Mirrors: Infinity, Unity, and Life in Leibniz's Philosophy, Chap. 1.

[8] Antognazza (2009), 167.

title).[9] Descartes further argues that, since we are finite beings, we should avoid discussing the infinite and thus avoid the paradoxes surrounding it.

> this is why we should not concern ourselves to respond to those who ask if half of an infinite line is infinite, and whether an infinite number is even or odd, and other similar things, because only those who imagine that their spirit is infinite have to examine such difficulties.[10]

Descartes expands on this point in a letter to Mersenne:

> I have read M. Morin's book. Its main fault is that he always discusses the infinite as if he had completely mastered it and could comprehend its properties. This is an almost universal fault which I have tried carefully to avoid. (28 Jan. 1641, Descartes To Mersenne; AT III 293; CSMK 171–172)[11]

According to Descartes, since our mind is finite, we should reserve the term "infinite" for God alone; for only God's nature can be properly called infinite. Other things that we perceive to have no limits, such as the extension of the universe, the number of the stars, or the division of matter should be regarded as indefinitely large or small, but not infinite. Descartes holds that the indefiniteness of such things does not belong to their nature but rather stems from the fact that human understanding is limited and deficient, and therefore cannot perceive the infinite (article 27).[12]

---

[9] For a translation of Leibniz's comments on this article, see LLC 25.

[10] «*c'est pourquoi nous ne nous soucierons pas de répondre à ceux qui demandent si la moitié d'une ligne infinie est infinie, et si le nombre infini est pair ou non pair, et autres choses semblables, à cause qu'il n'y a que ceux qui s'imaginent que leur esprit est infini qui semblent devoir examiner telles difficultés*» (article 26, CSM I 201–2; AT VIIIA 14–15).

[11] He adds: "I have never written about the infinite except to submit myself to it and not to determine what it is or what it is not." Descartes to Mersenne, 28 January 1641, AT III 293, translation in Ariew (1990), on p. 17.

[12] « 26. Qu'il ne faut point tâcher de comprendre l'infini mais seulement penser que tout ce en quoi nous ne trouvons aucunes bornes est. indéfini.

Ainsi nous ne nous embarrasserons jamais dans les disputes de l'infini; d'autant qu'il serait ridicule que nous, qui sommes finis, entreprissions d'en déterminer quelque chose, et par ce moyen le supposer ni en tâchant de le comprendre; c'est. pourquoi nous ne nous soucierons pas de répondre à ceux qui demandent si la moitié d'une ligne infinie est. infinie, et si le nombre infini est. pair ou non pair, et autres choses semblables, à cause qu'il n'y a que ceux qui s'imaginent que leur esprit est. infini qui semblent devoir examiner telles difficultés. Et, pour nous, en voyant des choses dans lesquelles, selon certains sens, nous ne remarquons point de limites, nous n'assurerons pas pour cela qu'elle soient infinies, mais nous les estimerons seulement indéfinies. Ainsi, parce que nous ne saurions imaginer une étendue si grande que nous ne concevions en même temps qu'il y en peut avoir une plus grande, nous dirons que l'étendue des choses possibles est. indéfinie; et parce qu'on ne saurait diviser un corps en des parties si petites que chacune de ses parties ne puisse être divisée en d'autres plus petites, nous penserons que la quantité peut être divisée en des parties dont le nombre est. indéfini; et parce que nous ne saurions imaginer tant d'étoiles que Dieu n'en puisse créer davantage, nous supposerons que leur nombre est. indéfini, et ainsi du reste.

27. Quelle différence il y a entre indéfini et infini.

Et nous appellerons ces choses indéfinies plutôt qu'infinies, afin de réserver à Dieu seul le nom d'infini; tant à cause que nous ne remarquons point de bornes en ses perfections, comme aussi à cause que nous sommes très assurés qu'il n y en peut avoir. Pour ce qui est. des autres choses, nous savons qu'elles ne sont pas ainsi absolument parfaites, parce qu'encore que nous y remarquions

Leibniz's note on Descartes's distinction (articles 26 and 27) in 1675 reads as follows:

> Instead of 'infinite', he recommends that we use the term 'indefinite', i.e. that whose limits cannot be found by us, and that the term "true infinity" should be reserved to God alone. But contrary to this, in Part 2, article 36, matter is admitted to be really divided by motion into parts that are smaller than any assignable, and therefore actually infinite. (A 6.3 214; LLC 25)

It is interesting to observe that already in his early "Theory of Abstract Motion" of 1671 Leibniz sharply objects to Descartes's distinction between the infinite and the indefinite with a similar line of reasoning, that is, by defending the actual divisibility of the continuum. He writes:

> There are actually parts in the continuum, contrary to what the most acute Thomas White believes, and *these are actually infinite*, for Descartes's "indefinite" is not in the thing, but in the thinker. (Winter 1670–1671, A 6.2 264; LLC 339)

From this note we learn that, even before he had direct access to Descartes's *Principles of Philosophy*, Leibniz criticized Descartes for grounding the distinction between the infinite and the indefinite epistemologically, that is, in human limitations for comprehending infinity. Both remarks (of 1671 and of 1675) suggest that Leibniz is not wary of endorsing the actual division of matter to infinity. Moreover, his 1675 remark suggests that he thinks that Descartes, too, presupposes something like the actual division of matter to infinity but uses the terminology of indefinite division instead. Leibniz refers to Part 2, article 36 of Descartes's *Principles*. Note, too, that Leibniz already recasts Descartes's position in his own terms: rather than referring to the "indefinitely divisible" he uses the phrase (which I suspect he adapts from Hobbes) "smaller than any assignable" – a phrase, which, for him, implies the syncategorematic sense of infinity. Leibniz's syncategorematic use of infinity implies that an infinite term would designate a finite term, larger or smaller than any assignable, so that one could vary for the sake of calculation. It does not imply that there is an infinitely small or infinitely large quantity.[13] Assuming his syncategorematic use of the infinite, Leibniz objects to Descartes's attempt to reserve the use of infinity for God alone. Instead, Leibniz would hold that there are many things (other than God) that can be adequately described as (either actually or potentially) infinite.

In spite of Leibniz's critique of Descartes's distinction between the infinite and the indefinite, there is one issue on which they agree: for both, the infinity of God is absolute, and, in the terms I will be using here, for both, the infinity of God is not seen as a quantitative kind of infinity. In other words, for both Leibniz and Descartes,

quelquefois des propriétés qui nous semblent n'avoir point de limites, nous ne laissons pas de connaître que cela procède du défaut de notre entendement, et non point de leur nature. »

[13] This view of infinity is developed and argued for by Richard Arthur. It is further developed in his chapter in this volume and I refer the reader to it.

the infinity of God does not relate to greatness in magnitude.[14] In a letter to Henry More, Descartes makes this point rather explicitly:

> God is the only thing I positively understand to be infinite. As to other things like the extension of the world and the number of parts into which matter is divisible, I confess I do not know whether they are absolutely infinite; I merely know that I know no end to them, and so, looking at them from my own point of view, I call them indefinite.[15]

In his second letter to More, Descartes writes:

> I say … that the world is indeterminate or indefinite, because I do not recognize in it any limits. But I dare not call it infinite as I perceive that God is greater than the world, not in respect to His extension, because, as I have already said, I do not acknowledge in God any proper [extension], but in respect to His perfection.[16]

The infinity of God, according to Descartes, relates primarily, and perhaps exclusively, to his perfection. With respect to the infinity of God, Leibniz's view is very similar to Descartes's. God's infinity does not pertain to extension or to any magnitude or other quantitative feature; rather, God's infinity pertains exclusively to his perfection.[17] It goes without saying that, for both Descartes and Leibniz, God is defined (in accordance with the tradition) as the most perfect being (*Ens Perfectissimum*). As Leibniz states, "The absolute is prior to the limited". "And just so the unbounded is prior to that which is has a boundary [terminus], since the boundary is something added" (A 6.3 502; A 6.3 392; GP I 224). As Robert Adams clarifies:

> Leibniz's conception of divine perfection commits him to agree with Descartes that, in its own nature, the divine infinity or perfection is primitive – that it is unanalyzable and not a negation of the finite. For him, as for Descartes, the infinite, in properties capable of infinity, is the primary case, and the finite is formed by limitation, or partial negation, of the infinite (NE 157f). (Adams 1994, p. 116)

But Leibniz's agreement with Descartes ends here. For Descartes states that there is no other thing that we should qualify as infinite – that we should refrain from ascribing infinity to things which seem unbounded to us because, in the final analysis, we cannot comprehend what infinity means. Further, Descartes argues that

---

[14] For more details on this, see Anat Shechtan's chapter in this volume.

[15] Descartes to Henry More, 5 February, 1649, CSMK 364.

[16] Descartes's second letter to Henry More, 15, May 1649, quoted in Koyré (1957, p. 122) (and not translated by CSMK). In a letter to Clerselier from 23 April 1649, Descartes explains, "[b]y 'infinite substance' I mean a substance which has actually infinite and immense, true and real perfections" (AT V 355; CSMK 377).

[17] It is arguable that perfection could still be measured in degrees, and that God's perfection could thus be subject to a quantitative conception of the infinite as, for example, that which is beyond all quantitative measure of perfection. As far as I can tell, Leibniz rejects this idea. In the first paragraph of the *Discourse on Metaphysics* he argues that, unlike magnitudes, perfection admits of a maximum. But this is precisely because perfection is not seen as a magnitude in the first place, and thus is not susceptible to the contradiction inflicting infinite quantities. As Reed Winegar points out, Kant seems to answer this question by denying that we have any standard for asserting the quantitative measure of a degree of perfection since all of our standards for quantitative measure are sensible and God's perfection is supersensible.

finite beings (such as we are) should not pretend to understand, or even try to understand, something infinite. Leibniz's attitude is almost the inverse. For Leibniz, there are many things (series, worlds, individuals) that can be (and, in fact, must be) understood as infinite – some as actually infinite, some as potentially infinite; some as quantitatively infinite, and some as qualitatively infinite (God). Thus, in describing all these things as infinite, Leibniz is using different senses of infinity. And, as noted above, he agrees with Descartes that only God is infinite in the *absolute* and non-quantitative sense.

One might argue that this disagreement between Leibniz and Descartes is merely about words, and that, in the end, there is no substantial difference in their positions. It is arguable that Leibniz's distinction between different senses of infinity comes down to something quite similar to Descartes's distinction between the infinite and the indefinite. Indeed, in his article, "Leibniz on the Indefinite as Infinite" (1998), Bradley Bassler argues that this is the case. According to Bassler, in Leibniz's notes from 1676, one finds a distinction that is very similar to Descartes's. Bassler observes (on p. 850) that Leibniz approves of the indefinite progression of natural numbers and rejects the notion of an infinite number of (finite) numbers. This observation has some basis in the texts, but Bassler's assimilation of Leibniz's notion of (the syncategorematic) infinite with Descartes's notion of the indefinite blurs some important differences between their views. Bassler argues that, in his later work, "Leibniz takes the indefinite as infinite" (p. 852). However, it is at least clear that Leibniz himself thought that his disagreement with Descartes was not merely terminological but rather substantial.

Notice first that the distinction Bassler is referring to is drawn within the realm of mathematics or more broadly in the realm of quantity. Here, Leibniz uses the notion of infinite number as an illustration of something impossible, for an infinite number cannot be conceived and thus has no consistent notion. Yet, Leibniz qualifies as infinite other things, which he does not consider impossible. An infinite series is an example. Leibniz sees an infinite series as consistent because he defines a series through its generation rule (or law of the series) and not as a sum of its constituents.[18]

Second, as already noted, Leibniz rejects Descartes's view that the distinction between infinite and indefinite is due to our incapability as finite minds to understand the infinite. One reason for this conviction is his mathematical work during his years in Paris. His work on infinite series and the calculus shows that he sees both the notion of infinitesimally small as well as that of infinitely large as mathematically manageable and indeed very useful. But it should be noted that Leibniz's response to Descartes precedes his development of the calculus. And so it seems that his early commitment to investigate the infinite does not depend on his mathematical work. In addition, Leibniz holds that, although we certainly do not fully

[18] Quite generally, the realm of mathematics and of all beings of reason such as concepts, relations, and possibilities is characterized by potential rather (than actual) infinity. Thus a line, for example, may be divided to infinity in whatever way one likes but an existing body is already divided to infinity.

comprehend the infinite, we can nevertheless demonstrate some things about it. This point is expressed in a letter to Malebranche from 1679:

> The number of all numbers implies a contradiction, which I show thus: to any number there is a corresponding number equal to its double. Therefore the number of all numbers is not greater than the number of even numbers, i.e. the whole is not greater than its part. It is no use responding that our finite mind cannot comprehend the infinite, for we can demonstrate something about what we do not comprehend. And here we comprehend at least the impossibility, if this only means that there is a certain whole which is not greater than its part.[19]

Leibniz's conclusion from the above reasoning is that an infinite sum of parts, seen as a whole, is an impossible notion. But this negative result has some positive implications: it leads Leibniz, in contrast to Descartes, to make positive observations about the infinite. According to Leibniz, we can say that there are infinitely many things or parts of matter as long as we do not see them as a single whole or as a true unity. As early as 1672, Leibniz observed that "[t]here is no maximum in things, or what is the same, the infinite number of all unities is not one whole, but is comparable to nothing" (A 6.3 98; LLC 13).[20] Thus, for Leibniz, it would be misguided to reduce infinity to something that we call undefined or undetermined because we cannot comprehend it. Indeed, for Leibniz, there is no categorematic infinity of things. But the notion of infinity is extremely useful. As he notes in his piece from 1676, "Infinite Numbers":

> we conclude finally that there is no infinite multiplicity, from which it will follow that there is not an infinity of things either. Or it must be said that an infinity of things is not one whole, i.e. that there is no aggregate of them.[21]

Leibniz's conclusion here is that one can talk about infinitely many things as long as one does not regard these things as a totality or as making up a single whole

[19] Leibniz to Malebranche, 22 June 1679; GP I 338, translation in Brown (2005), on p. 479. This point comes up in other passages as well: "At last a certain new and unexpected light shined from where I least expected it, namely, from mathematical considerations on the nature of infinity. For there are two labyrinths of the human mind, one concerning the composition of the continuum, and the other concerning the nature of freedom, and they arise from the same source, infinity. That same distinguished philosopher I cited a short while ago preferred to slash through both of these knots with a sword since he either could not solve the problems, or did not want to reveal his view. For in his *Principles of Philosophy* I, art. 40–41, he says that he can easily become entangled in enormous difficulties if we try to reconcile God's preordination with freedom of the will; but, he says, we must refrain from discussing these matters, since we cannot comprehend God's nature. And also, in *Principles of Philosophy* II, art. 35, he says that we should not doubt the infinite divisibility of matter even if we cannot grasp it. But this is not satisfactory, for it is one thing for us not to comprehend something, and quite something else for us to comprehend that it is contradictory" (1689? "On Freedom"; AG 95). See also: "having contented himself with saying that matter is actually divided into parts smaller than all those we can possible conceive, [Descartes] warns that the things he thinks he has demonstrated ought not to be denied to exist, even if our finite mind cannot grasp how they occur. But it is one thing to explain how something occurs, and another to satisfy the objection and avoid absurdity." ("Pacidius to Philalethes", 29 Oct.–10 Nov. 1676, A 6.3 554; LLC 183–185)

[20] See also NE 2.17.

[21] April 10, 1676, "Infinite Numbers", A 6.3 503; LLC 101.

(that would also admit of parts).[22] This is an important point that Leibniz firmly holds for the rest of his career. As an example, consider this passage from Leibniz's letter to Bernoulli of 1699:

> Given infinitely many terms, it does not follow that there must be an infinitesimal term… I concede the infinite multiplicity of terms, but this multiplicity does not constitute a number or a single whole. It signifies nothing but that there are more terms than can be designated by a number. Just so, there is a multiplicity or complex of numbers, but this multiplicity is not a number or a single whole.[23]

As we have already seen earlier, Leibniz denies the possibility of infinite quantities. But this, he thinks, need not prevent us from using infinity. For to refer to infinitesimals or infinite series is not to refer to true wholes or to true entities. In other words, one need not suppose the existence of an infinitely small (or large) quantity (or entity) in order to use the infinitesimal calculus (or to apply infinity more generally).[24] Thus, for Leibniz, there is an actual infinity of real things or substances, and matter is actually infinitely divided. However, this actual infinity cannot, strictly speaking, be referred to as one or as a whole. Likewise, the concept of an infinitesimal is but a useful fiction.

Since Leibniz's critical response to Descartes's distinction between the infinite and the indefinite appears in early in his work (1671), it seems that Leibniz held this approach even before he started his serious work in mathematics (under the guidance of Huygens in Paris). Thus, it is possible that Leibniz's early comment points to some of the intuitions that might have led him to *develop* his calculus.[25] Be this as it may, my main point here is that Leibniz's approach to investigating the infinite stands in stark contrast to Descartes's. Descartes recommends avoiding any discussion of the infinite and, especially, pretending that we can comprehend it. Leibniz's attitude towards the question of infinity could not be more different. As we shall see, unlike Descartes, Leibniz does attempt to provide a positive account of the infinite. The productive ways in which infinity can be used in mathematics (as the development of the calculus makes evident) as well as in metaphysics (in using it to characterize individuals, worlds, etc.) will begin to unfold in light of his response to Galileo.

9.3 Leibniz's Encounter with Galileo

Between the autumn of 1672 and early 1673, while occupied with many other things, Leibniz found some time to read Galileo's *Discourses and Mathematical Demonstrations Concerning the Two New Sciences* (Antognazza 2009, 144). In this work Galileo presents several paradoxes concerning the infinite. He opens the

---

[22] See also the "Conversation of Philarete and Ariste" in AG 267; GP VI 592.

[23] 21 Feb. 1699, Leibniz to Bernoulli; GM III 575; L 514, translation is from Levey (1999), on p. 139.

[24] See Richard Arthur's discussion in his contribution to this volume.

[25] See Arthur in LLC xxxiii.

discussion with a geometrical example[26] and then presents an arithmetical argument. This argument, which Leibniz cites in several places,[27] is referred to in the recent literature as "Galileo's paradox". It runs as follows: "all numbers, comprising of the squares and the non-squares, are greater than the squares alone" (EN 78; LLC 356). In other words, the series 1, 2, 3, 4, 5, 6, … (of squares and non-squares) has members that the series of squares alone 1, 4, 9, 16, … does not have (2, 3, 5, 6,). But, "there are as many square numbers as there are their own roots, since every square has its own root, and every root its own square…. But, if I were to ask how many roots there are, it cannot be denied that there are as many as all the numbers, because there is no number that is not a root of some square. That being so, it must be said that the square numbers are as many as all the numbers, because they are as many as their roots, and all numbers are roots" (EN 78; LLC 356). On the one hand, there appear to be more numbers than squares, but, on the other hand, there are as many numbers as squares. Thus, it turns out that the quantity of squares is both "less than" and "equal to" the quantity of all numbers – a paradox. Given this paradox, one might be inclined to infer that the relations of "greater than", "less than", and "equal to" do not apply in the context of infinity. This is precisely Galileo's conclusion.

The radical conclusion Galileo draws from this argument is reiterated by Salviati: "from your ingenious argument we are led to conclude that the attributes 'larger,' 'smaller,' and 'equal' have no place either in comparing infinite quantities with each other or in comparing infinite with finite quantities" (EN 80). As Galileo writes, "I believe that these attributes of greatness, smallness, and equality do not befit infinities, about which it cannot be said that one is greater than, smaller than, or equal to one another" (EN 77–78; LLC 355).

Galileo concludes that insurmountable paradoxes arise when the notion of infinity is regarded as a quantity. The paradoxes he points to, show that the most basic properties that must pertain to a quantity (such as "bigger than", "smaller than", or "equal to") do not hold in the case of "infinite quantity" (EN 80). As Knobloch has stressed, if this is the case, infinities should not be regarded as *quanta* at all (Knobloch 1999).

Leibniz's conclusion from his readings of Galileo's *Two New Sciences* in 1672–1673, however, was rather different. Instead of determining that the infinite does not belong in the realm of quantity, Leibniz comes to the conclusion that the notion of *an infinite number, seen as a whole*, is impossible. It is impossible precisely because such a notion violates the axioms that govern the realm of quantity, more specifically, the axiom stating that the whole is greater than its part.[28] The following note

---

[26] This argument is sometimes referred to as *rota Aristotelis*. See EN 68, LLC 432.

[27] E.g. "Pacidius to Philalethes"; LLC 179.

[28] "[J]ust as the proposition 'the whole is greater than the part' is the basis of arithmetic and geometry, i.e., of the sciences of quantity, similarly, the proposition 'nothing exists without reason' is the foundation of physics and morality, i.e., the sciences of quality, or, what is the same (for quality is nothing but the power of acting and being acted on) the sciences of action, including thought and action" (*Confessio* 35; A 6.3 118). See also Knobloch (1999), p. 94. The notion of a whole is thus used in a technical sense so that notions of an infinite number and an infinite whole are equivalent. It hardly needs saying that this is not obvious. As Reed Winegar comments, Cantor would reject it

(on Galileo's *Two New Sciences*) summarizes Leibniz's response to Galileo's paradox:

> Among numbers there are infinite roots, infinite squares, infinite cubes. Moreover, there are as many square numbers as there are numbers in the universe. Which is impossible. Hence it follows either that in the infinite the whole is not greater than the part, which is the opinion of Galileo and Gregory of St. Vincent, and which I cannot accept; or that infinity itself is nothing, i.e. that it is not one and not a whole. (Fall 1672, A 6.3 168; LLC 9)

Leibniz argues that one cannot accept the result that the series of natural numbers is equal to the series of their squares; for, if this were permitted, the whole (the series of natural numbers) would not be greater than its parts (the series of squares). However, he finds "it difficult to agree" with Galileo's conclusion that the "appellations of greater, equal, and less have no place in the infinite" (A 6.3 551; LLC 179), "[f]or who would deny that number of square numbers are contained in the number of all numbers. But to be contained in something is certainly to be a part of it, and I believe it to be no less true in the infinite than in the finite that the part is less than the whole" (A 6.3 551; LLC 179).

Richard Arthur nicely presents the choices Leibniz sees as emerging from Galileo's paradox in the form of the following dilemma:

> [Leibniz] identifies two candidates for rejection: (W) that in the infinite the whole is greater than the part, and (C) that an infinite collection (such as the set of all numbers) is a whole or unity … Leibniz upholds W, and this leads him to reject C. Cantor upholds C, and this leads him to reject W. (Arthur 2001, pp. 103–104).

Leibniz concludes that the whole is greater than the parts even for the infinite, and therefore must deny that an infinite plurality can be regarded as a whole. Thus, according to Leibniz, there cannot be a number of all numbers, or an infinite number. This implies that an infinite collection of elements cannot be regarded as a genuine unit.

As noted, the conclusion Galileo draws from his paradoxes is that, if the most basic relations of quantity do not hold in the realm of infinity, then infinity cannot be regarded as a quantity. This implies that the finite and the infinite belong to different categories that cannot even be compared. Eberhard Knobloch (1999) has put this point as follows: according to Galileo, "[A]n 'infinite quantity' would … be a 'contradiction in terms', because an infinite would lack precisely those properties that characterize a quantity" (p. 94).

Knobloch further maintains that Leibniz's response to Galileo was to show, through his mathematical work on the calculus, that infinities can be handled in quantitative and precise terms. However, as I will argue below, while Knobloch's account is correct, it leaves out an important part of the story. Knobloch observes that Leibniz's calculus and his interpretation thereof show how the infinite can be dealt with mathematically, i.e., as a quantity. This is no doubt true. However, this

---

on the grounds that there is no infinite number but a set is an infinite "whole". Kant would also reject it. He denies that there is an infinite number, but he is happy to refer to infinite magnitudes as wholes (wholes that are beyond measure).

observation holds only for one sense of infinity, which Leibniz (not surprisingly) reserves for quantities, numbers, and magnitudes (and for which he develops his syncategorematic approach). But there is another sense of infinity for which Leibniz actually accepts Galileo's position that the infinite cannot be regarded as a quantity. This is the notion of infinity that he would apply to being in general, and, as we have seen in the previous section, to God's being (and perfection) in particular. I suggest that this distinction, between a quantitative and a non-quantitative sense of infinity, is of great consequence not only for Leibniz's resolution of the paradoxes of infinity but also plays an important role in his metaphysics.[29]

To a large extent, Leibniz's approach to infinity can be seen as a complex response to, and a sophisticated development of, Galileo's conclusions. In working his way through Galileo's *Dialogues*, Leibniz has "acquired" two challenges. (i) On the one hand, his recognition of Galileo's paradox motivates him to distinguish between a kind of infinity that he regards as non-quantitative and applicable to beings (and especially to the most perfect Being) and a kind of infinity that is quantitative. The latter kind of infinity is applicable to the mathematical domain, as well as to quantitative features of the metaphysical domain (such as the division of matter, for example).[30] (ii) This sets another challenge for Leibniz: to show how one can treat the notion of infinity within mathematics, that is, in a quantitative sense. Much of Leibniz's work on infinite series and the calculus during his Paris years can be seen as a response to this task. One result of his efforts is the syncategorematic interpretation of infinite terms (seen as useful fictions) that is presented and defended in detail by Richard Arthur.[31]

At the same time, Leibniz's resolution of Galileo's paradox in terms of rejecting infinite number gives rise to another major problem. The problem is how to account for the difference between the notion of an infinite number (which he regards as impossible) and the notion of an infinite being (the primary and most perfect being), which he regards as possible; in fact, not only as possible but also as implying a necessary being – one whose non-existence is impossible.

Leibniz's claim that "the number of all numbers is a contradiction" (e.g., A 6.3 463; DSR 7) appears in his Paris notes from 1675 to 1676, a period in which he was developing his views about infinity in various domains. At the same time, Leibniz was also engaged, among many other projects, in distinguishing between possible and impossible notions. Indeed, Leibniz's view of possibility plays a central role in

---

[29] For an elaboration of this point see my book *Living Mirrors: Infinity, Unity, and Life in Leibniz's Philosophy*, forthcoming with Oxford University Press.

[30] On this point, see Richard Arthur's contribution to this volume. Arthur defends the position that the syncategorematic reading of infinity applies both to mathematics and metaphysics. Thanks to this reading, Leibniz can make claims such as there are infinitely many (understood syncategorematically) monads, while rejecting infinite number.

[31] See his introduction to LLC and Arthur (2013), as well as his contribution to the present volume.

his metaphysics.[32] As early as his "Confession of a philosopher" (1672–1673), Leibniz defines a possibility such that *x* is possible if it has a notion whose internal constituents are consistent. It is important to observe that, in this context, Leibniz is using the notions of the "number of all numbers" (*numerum omnium numerorum*) and the "greatest or maximal number" (*numerus maximus*) as an illustration of an impossible notion, that is, a notion whose internal constituents imply a contradiction. In the same texts, Leibniz also uses the notion of the number of all numbers in contrast to that of "the greatest or the most perfect being" (A 6.3 572; DSR 91).

Comparing the notion of the greatest being with the notion of the greatest number gives rise to a severe problem. Leibniz states this problem in a letter to Oldenburg from December 1675:

> Whatever the conclusions which the Scholastics ... and others derived from the concept of that being whose essence is to exist, they remain weak as long as it is not established whether such being is possible, provided it can be thought. To assert such a thing is easy; to understand it is not so easy. Assuming that such a being is possible or that there is some idea corresponding to these words, it certainly follows that such a being exists. But we believe that we are thinking of many things (though confusedly) which nevertheless imply a contradiction; for example, the number of all numbers. We ought strongly to suspect the concepts of infinity, of maximum and minimum, of the most perfect, and of allness (*omnia*) itself. Nor ought we believe in such concepts until they have been tested by that criterion which must, I believe, be credited to me, and which renders truth stable, visible and irresistible. (GM I 83-84; L 257)

The worry raised by Leibniz here is made even clearer in a letter to Elizabeth, written 3 years later (1678). There, Leibniz considers several examples of impossible notions (such as those of the squared circle and of the greatest speed) and writes:

> we think about this greatest speed, something that has no idea since it is impossible. Similarly, the greatest circle of all is an impossible thing, and the number of all possible units is no less so; we have a demonstration of this. And nevertheless, we think about all this. That is why there are surely grounds for wondering whether we should be careful about the idea of the greatest of all beings, and whether it might not contain a contradiction. (A 2.1 433–38; AG 238)

Leibniz's reasoning here is very clear. Since we entertain thoughts about things such as the greatest speed and the greatest number, which upon analysis prove to be contradictory, we ought to examine whether the idea of the greatest of all beings might not turn out to be contradictory as well. In fact, we often use concatenations of words that do not correspond to any idea and which might well turn out to be contradictory.

Leibniz's problem is therefore to show that, while the greatest number is contradictory and thus impossible, the greatest or most perfect being, i.e., God, is not. This problem, however, is central for understanding Leibniz's complex approach to infinity and the context in which it develops. The formulation of this dilemma in this section sets the stage for exploring Leibniz's solution to this problem through his encounter with Spinoza.

---

[32] For more details regarding Leibniz's view of possibility, see Nachtomy (2007).

9.4 Leibniz's Encounter with Spinoza

In his years in Paris, and particularly in 1675 and 1676, Leibniz shows great curiosity and interest in Spinoza's philosophy, which he often discusses with their common friend Tschirnhaus. While he fails to obtain the unpublished manuscript of the *Ethics* held by Tschirnhaus, he does get to read Spinoza's letter to L. Meyer in which Spinoza lays out his views on infinity (Ep. 12). Leibniz read and annotated this letter in April 1676.[33]

In 1676, Leibniz reluctantly traveled from Paris back to Hanover and he made a point to travel via The Hague in order to meet Spinoza. The two philosophers met, but this was their only meeting. Their philosophical systems, however, have many more meeting points. In The Hague, Leibniz showed Spinoza his amended version of Descartes's proof for the existence of God. According to this version of the argument, the notion of the *Ens Perfectissimum* entails existence; for it includes all perfections, and existence is considered a perfection. Leibniz found this reasoning unsatisfactory for the following reason: one needs to show not only that the conclusion follows from the premises, but also that the definition of the *Ens Perfectissimum* as the subject of all perfections is consistent – a point that was taken for granted by all previous upholders of the argument. In other words, Leibniz argues that, in order to prove that a most perfect being exists, one has to show first that this notion is consistent (A 6.3 572; DSR 91; A 6.3 583; DSR 105-07).[34]

The issue Spinoza and Leibniz discussed in their meeting in The Hague is highly indicative of some of the remarkable affinities, as well as some of the deep rifts, between their views regarding the nature of the infinite, and especially the relation between the notions of infinite being and infinite number. While Leibniz's approach implies that the existence of an infinite and most perfect being follows from its essence, Spinoza holds that, since being finite involves some negation, infinity expresses (or, one might say, is) the very absolute affirmation of existence.[35]

---

[33] See Lærke (2008). This letter is not Leibniz's only source regarding Spinoza's views at the time. He receives quite accurate information on the *Ethics* from Tschirnhaus, with whom he discusses Spinoza's metaphysics as well as questions of mathematics (see for example Leibniz's letter of May 1678, GM IV 451–63; L 294–99).

[34] Cf. a note from 1676, in which Leibniz writes, "[i]n the chapter of St. Thomas' *Summa Contra Gentiles* which is entitled 'Whether the existence of God is known *per se*,' there is a reference to an elegant argument which some use to prove the existence of God. The argument is: God is that than which nothing greater can be thought. But that than which nothing greater can be thought cannot not exist. For then some other thing, which cannot not exist, would be greater than it. Therefore God cannot not exist. This argument comes to the same as one which has often been advanced by others: namely, that a most perfect being exists. St. Thomas offers a refutation of this argument, but I think that it is not to be refuted, but that it needs supplementations. For it assumes that a being which cannot not exist, and also a greatest or most perfect being, is possible" ("On Truths, the Mind, God, and the Universe" 15 April, 1676, A 6.3 510–511; DSR 63).

[35] "Since to be finite is some negation and to be infinite is an absolute affirmation of the existence of some nature, it therefore follows from proposition 7 that any substance must necessarily exist" (EIP8S). Spinoza argues further that one can adequately consider the uniqueness of being (i.e., of substance) with respect to its existence alone and not its essence. As Spinoza writes to Jarig Jelles,

Another crucial difference in their metaphysical systems is that, for Spinoza, there is only a single substance, whereas Leibniz speaks of infinitely many beings. As it is usually put, Spinoza is a substance monist and Leibniz a substance pluralist. Indeed, this distinction captures a major difference between Spinoza's and Leibniz's metaphysical systems, and yet a close analysis reveals a fundamental agreement regarding the claim that any substance is, by definition, both infinite and unique. In light of this similarity, I will suggest that, when Leibniz and Spinoza say that the divine substance is infinite, it is primarily understood in a non-quantitative sense along the lines observed in Leibniz's encounter with Descartes and Galileo.[36]

9.5 Leibniz Reads Spinoza's Letter on the Infinite

Like Leibniz, Spinoza has to account for the difference between the infinity of number and the infinity of God. Spinoza explicitly defines God as "a being absolutely infinite, that is, a substance consisting of an infinity of attributes" (EID6).[37] To the definition of God as an absolutely infinite being Spinoza adds the following explanation: "I say absolutely infinite, not infinite after its kind: for, of a thing infinite only after its kind, infinite attributes may be denied; but that which is absolutely infinite, contains in its essence whatever expresses reality, and involves no negation" (ibid). But if the infinity of the attributes is understood in numerical terms, the definition of God would seem susceptible to the same problem we have observed in Leibniz, viz., the tension between infinite number and infinite being. In letter 12, Spinoza takes the following approach to this problem: he distinguishes between

"in an *Appendix to the Principles of Descartes, Geometrically Demonstrated* I established that God can be called one [*unum*] or unique [*uniqum*] only in a very inappropriate sense, I respond that a thing cannot be called one and unique with respect to essence but only with respect to existence. We conceive of things as existing in a certain number of exemplars only if they are brought under a common genus" (Ep. 50 to Jarig Jelles, Gebhardt IV 239). I take this to imply that, according to Spinoza, one cannot *conceive* of the unique and infinite being in abstraction from its existence, as a pure essence. In addition, one may talk about the unique existing thing in a numerical sense only in an inappropriate sense. For the category of number can only apply to things that can be "brought under a common genus", which obviously does not hold of a unique being.

36 The thesis of the non-quantitative sense of infinity, as presented below, also provides a partial explanation for why Leibniz was attracted to Spinoza's philosophy during his years in Paris (especially in 1675–1676) and, at the same time, why he ultimately moved away from it while reading and commenting on Spinoza's *Ethics* in 1678. For details of the complex way in which Leibniz read Spinoza, see Lærke (2008).

37 The translation is significant here. It can also be translated "consisting of infinite attributes". This is important in identifying the kind of infinity that is at work here. In addition, this plays into the debate regarding the number of attributes that God is said to have. Those who translate it as "an infinity" tend to hold that there is a numeric infinity of attributes, while "infinite attributes" is related to the infinite nature of the attributes. I prefer the former but have opted to use Curely's translation throughout.

different kinds of infinity. In particular, he distinguishes between a kind of infinity that applies to the unique indivisible substance and a kind of infinity that applies to divisible quantities (See LLC 103–105; Curley 202). This approach emerges explicitly when Spinoza discusses the nature of the infinite and how to dissolve the traditional paradoxes surrounding it.[38]

Spinoza's argumentation in this letter is of considerable complexity. However, one point that does seem rather clear is that the notion of infinity that may apply to the substance is non-quantitative. Since Spinoza identifies God with the unique substance, God's infinity is not comparable to that of numbers. According to Spinoza, any reference to number presupposes a limitation and would imply that it is finite. Indeed, the tendency to describe a substance with numerical infinity is misguided and generates contradictions. The way out from the contradictions affecting the infinite is to avoid the common confusion between the *quantitative sense* of infinity that can adequately be ascribed to numbers and divisible quantities, and the sense of infinity that can adequately be ascribed to a unique and indivisible substance.

It is worth noting that Leibniz begins his annotations on Spinoza's letter by stating that Spinoza "demonstrates that every substance is infinite, indivisible, and unique" (A 6.3 275; LLC 101). Leibniz then copies (almost to the letter) Spinoza's definitions of substance (EID3) and of God (EID6). This is certainly indicative of the interest Leibniz takes in reading this letter. Particularly instructive here is Leibniz's addition to Spinoza's definition of God.

> *He defines God as follows:* that which is an *absolutely* infinite being, i.e. a substance consisting of infinite attributes, each of which expresses an infinite and eternal essence and is thus immense [*immensum*]. (LLC 103)[39]

The clause "*adeoque immensum est*" is not part of Spinoza's definition but is added by Leibniz. In his annotations to this letter (A 6.3 282; L 24; LLC 115), Leibniz states the following: "I have always distinguished the *Immensum* from the *Interminato*, i.e., that which has no bound [*seu terminum non habente*]." In notes from this period, Leibniz is using *Immensum* as a noun – the *Immensum* – designating God as infinite but without measure. He also uses *Immensum* as "that which persists during continuous change in space … and is one and indivisible" (A 6.3 519, see LLC 450). Evidently, Leibniz is using the notion of *Immensum* in more than one sense. Likewise, divine immensity is taken as the "basis of space" (ibid). However, it seems that, unlike the current English connotations of the word "immense", Leibniz does not use *immensum* here to indicate a large or immense *magnitude;* rather, he uses it in a way much closer to its literal meaning in Latin, that

[38] In his exposition of Descartes's *Principles of Philosophy*, Spinoza mentions some of the traditional difficulties associated with infinity: "if an infinite is not greater than another, quantity A will be equal to its double, which is absurd"; "whether half an infinite number is also infinite, whether it is even or odd, and the like" (Gebhardt I 190; see also 192–196).

[39] «Deum sic definit. Quod sit Ens absolute infinitum, hoc est. substantia constans infinitis attributis, quorum unumquodque infinitam et aeternam essentiam exprimit adeoque immensum est. » (A 6.3 276).

is, to indicate something *beyond any measure*, or more precisely, something that has no measure (and is therefore impossible to measure) – something that *cannot* be measured because it does not belong to the category of quantity.[40]

Both recent English translators, Parkinson and Arthur, have emphasized this point (see DSR 122, n. 92 and LLC 450). To avoid the current English connotations of "immense", Parkinson renders *immensum* as "immeasurable", so that the Latin negation of measure (*mensura*) is conspicuous in the translation. As Arthur notes in the Glossary to his edition, "*Immensum* can be synonymous with 'infinite' or 'beyond measure', as Leibniz employs it in Aiii4: 95; and at Aiii60: 475, where Leibniz distinguishes this species of the infinite from the unbounded" (LLC 450, Latin–English Glossary).

Thus, when Leibniz adds his gloss to Spinoza's definition of God, i.e., that the absolutely infinite being is also *immensum*, he refers to one of his own notions of infinity, viz., that which is beyond measure. It is in this sense that *immensum* is distinguished from the unbounded. The unbounded infinite designates a *measurable* quantity, whereas *immensum* designates something that *cannot be measured*. Thus, Leibniz wishes to emphasize that the infinity of God is beyond any measure – and therefore cannot to be described in measurable or quantitative terms. The main point here is that Leibniz's addition indicates that, in his eyes, the infinity of the divine substance cannot be quantified or measured but rather belongs to an altogether different category.[41]

Leibniz then adds a very interesting note on a being conceived through itself (*per se concipi*):

> we understand through itself only that which is its own cause, i.e., that which is necessary, i.e., is a being in itself. And so it can be concluded from this that if we understood a necessary being, we would understand it through itself. But it can be doubted whether we do understand a necessary being, or, indeed, whether it could be understood [*intelligatur*] even if it were known or recognized [*cognosci*]. (A 6.3 275; LLC 101)

In reading Spinoza's letter, Leibniz seems to recall the difficulty of showing that the notion of a necessary being can be understood or, in other words, that it is intelligible. According to Leibniz, in order to show that something is intelligible, one has to show that its concept is consistent. Thus, it seems that, in reading Spinoza's letter, Leibniz is still occupied with his own problem.

Given Leibniz's preoccupation with the tension between the possibility of an infinite being and the impossibility of an infinite number, it is not surprising that he

---

[40] This is consistent with the view of absolute perfection. Perfection does not belong to the category of quantity/magnitude in the sense that perfection may admit of degrees but God's perfection is of a magnitude beyond measure because it is maximally perfect and there is no maximum quantitative degree. For a clear expression of this point, see the first paragraph of the *Discourse on Metaphysics* (AG 35; A 6.4 1531). See also Leibniz's direct reference to immensity in the *New Essays on Human Understanding*, where he writes: "I believe that we have a positive idea of each of these [infinity, eternity, and immensity]. This idea will be true provided that it is conceived not as infinite whole but rather as an absolute, i.e., an attribute with no limits" (NE II xvii; B&R 160).

[41] For a slightly different emphasis on Leibniz's addition of the word *immensum*, see Lærke (2008, pp. 469–477 and pp. 424–25).

is interested in the way that Spinoza connects the definitions of substance, God, and infinity. He agrees with Spinoza that any substance "is infinite, indivisible, and unique." Yet, according to him, the possibility of such a being needs to be demonstrated. As we saw in the previous section, Leibniz argues that an infinite collection of discrete units is impossible and cannot be regarded as a whole. At the same time, he clearly regards God as an infinite unity. In fact, he calls God the one-all (*unus omnia,* A 6.3 385), and maintains that such a being is possible.

In light of this, one can reasonably suppose that Leibniz would seek support for his line of reasoning regarding the possibility of an infinite being and the impossibility of an infinite number. Such support may indeed be found in Spinoza's letter. Towards the beginning, Spinoza notes:

> everyone has always found the problem of the Infinite very difficult. Indeed insoluble.[42] This is because they have not distinguished between what is infinite as a consequence of its own nature, or by the force of its definition, and what has no bounds, not indeed by the force of its essence, but by the force of its cause. And also because they have not distinguished between what is called infinite because it has no limits and that whose parts we cannot explain or equate with any number, though we know its maximum and minimum. Finally, they have not distinguished between what we can only understand, but not imagine, and what we can also imagine. (LLC 103; Curley 201)

He adds:

> If they have attended to these distinctions, I maintain that they would never have been overwhelmed by such a great crowd of difficulties. For then they would have understood clearly what kind of Infinite cannot be divided into any parts, or cannot have any parts, and what kind of Infinite can, on the other hand, be divided into parts without contradiction. They would also have understood what kind of Infinite can be conceived to be greater than another Infinite, without any contradiction, and what kind cannot be so conceived. (LLC 103–105; Curley 202)

By means of these distinctions, Spinoza qualifies and restricts the way in which infinity can be ascribed to substance.[43] Spinoza's distinction thus suggests an attractive approach to Leibniz's problem. According to Spinoza, one kind of infinity (the one pertaining to infinite being) "cannot be divided into any parts, or cannot have any parts", and the other kind of infinity (the one pertaining to modes) "can... be so divided into parts without contradiction." This is of course related to Spinoza's view that, strictly speaking, a substance is infinite and indivisible (EIP15S). For this reason, a substance is not divided into parts; rather, its attributes have various modes.

According to Spinoza, the kind of infinity that we can ascribe to substance is such that "we cannot explain or equate with any number". An infinite substance, on this view, is non-divisible and cannot be understood in numerical terms. For this reason, the use of this kind of infinity would not involve the contradictions that affect things whose enumeration requires comparison and abstraction by the imagi-

[42] See Ariew (1990, pp. 16–31).

[43] For Spinoza, these different kinds of infinity correspond to different kinds of things, viz., substance, attributes, and modes. Duration, number, and motion are seen as mere auxiliaries of the imagination, which serve as measures of divisible magnitudes. Cf. Gueroult (1968), Ariew (1990), as well as Lærke (2008).

nation. In fact, Spinoza maintains that enumeration involves abstraction and comparison of things under a common genus by means of the imagination.[44] However, the kind of infinity that pertains to a substance cannot even be adequately conceived by the imagination but only by the intellect. This can clearly be seen in the following passage from Ep. 12 (which summarizes EIP15S):

> we conceive quantity in two ways: either abstractly, or superficially, as we have it in the imagination with the aid of the senses; or as substance, which is done by the intellect alone. So if we attend to quantity as it is in the imagination, which is what we do most often and most easily, we find it to be divisible, finite, composed of parts, and one of many. But if we attend to it as it is in the intellect, and perceive the thing as it is in itself, which is very difficult, then we find it to be infinite, indivisible and unique, as I have already demonstrated sufficiently to you before now. (A 6.3 278; LLC 107; Curley 202–203)[45]

This last point – which Leibniz mentions in the first line of his annotations to Spinoza's letter – suggests a way out of the inconsistency Leibniz identifies in infinite quantity. In line with Spinoza's reasoning, Leibniz can distinguish between "beings" and "non-beings" by observing that each applies to a different kind of infinity. And this would account for his regarding an infinite being as possible and an infinite number as impossible. At the same time, the term 'infinite' in the *concept* of an infinite being, if it refers to the infinity of attributes, is to be used in a syncategorematic sense.

Thus, what's most pertinent for Leibniz's purposes in Spinoza's letter can be paraphrased as follows: any number is by definition limited. For this reason, it is also measurable. By contrast, God's infinity cannot be quantified, measured, or numbered, precisely because this would imply limiting it (or seeing it as limited), as well as viewing God as a divisible and discrete entity, which Spinoza clearly regards as absurd. This suggests that, for Spinoza, "infinity" is used differently when ascribed to numbers (or more generally to divisible and discrete quantities, or to a feature of modes and abstractions) and when it applies to the all-inclusive substance or God. A substance is said to be infinite on account of its completeness and absolute perfection. Therefore, for Spinoza, it must be *in*divisible and admit of no parts. In this sense, a substance is infinite in a non-quantitative sense.

Given the context of Leibniz's problem presented in the previous section, it should now become clear why Leibniz would be receptive to such a view. Indeed, he seems to agree with Spinoza's analysis. Yet, as is typical of him, Leibniz does not simply accept Spinoza's analysis; rather, he recasts Spinoza's distinction in his own terms and appropriates it for his purposes.[46] In his annotations, he writes,

[44] See EIP15, Ep. 34, Ep. 50, and the next section for more details.

[45] Cf.: "If therefore we consider quantity as it is in the imagination, that which is the most ordinary, we find that it is finite, divisible and composed of parts; if, on the contrary, we consider it as it is in the understanding and we conceive it insofar as it is substance, then, as we sufficiently demonstrated, we will find it to be infinite, unique and indivisible" (EIP15S).

[46] Obviously, I do not argue here for a direct influence in the sense that Spinoza's letter is the exclusive or even the main source for Leibniz's views on infinity. Rather, I claim that Leibniz's attraction to Spinoza's view is evident in his annotations and that his response to Spinoza's views is revealing of and serves him to articulate his own views.

> I set in order of degree: *Omnia*; *Maximum*; *Infinitum*. Whatever contains *everything* is maximum in entity; just as a space unbounded in every direction is maximum in extension. Likewise, that which contains everything is most infinite [*infinitissimum*], as I am accustomed to call it, or the absolutely infinite. The *Maximum* is everything of its kind, i.e., that to which nothing can be added, for instance, a line unbounded on both sides, which is obviously also infinite; for it contains every length. Finally those things are *infinite in the lowest degree* whose magnitude is greater than we can expound by an assignable ratio to sensible things, even though there exists something greater than these things... For a maximum does not apply in the case of numbers. (A 6.3 282; LLC 114–15)[47]

As Lærke notes,

> If one compares this classification with the one proposed in Spinoza's letter 12, one is struck by their similarity. First, the distinction between *maximum* and *omnia* evokes the distinction between the attributes, which are infinite "in their kind" in EIdef.4 and the 'absolutely infinite' substance in EIdef.6 – which is exactly the definition reproduced at the beginning of the *Communicata ex litters domini Schulleri*. [Likewise] there is a strong resemblance between that which Leibniz calls '*immensum*' and that which Spinoza calls 'infinite by nature'. (Lærke 2008, 433, my translation)

The similarity between *immensum* and "infinite by nature" is particularly remarkable. As Lærke also notes, "that which is infinite by nature or by virtue of its definition, is the substance" (ibid, 430). We have already seen that Leibniz amends Spinoza's definition of God with the clause "that which is *immensum*", and thus it is very natural to see this highest degree as applying to God. The lowest degree of infinity calls to mind Leibniz's syncategorematic notion of infinity and thus things that can be quantified. For lack of space, I do not discuss the middle degree here.

Still, there are some significant differences between Leibniz and Spinoza here. The most conspicuous difference is terminological: Leibniz reformulates Spinoza's distinction in terms of degrees. *Omnia*, he says, "is the highest degree, [it] is everything, and this kind of infinite is God, since he is all one; for in him are contained the requisites for existing of all the others" (A 6. 3 385; LLC, 43). Elsewhere, and later in his career, Leibniz is also very clear that the highest degree, the "absolutely infinite", applies to God alone. For example, in a letter to Des Bosses, from 11 March 1706, he notes that "only indivisible and absolute infinite has true unity: it is God" (GP II 305). In the *New Essays* (2.17.1), he writes that "rigorously speaking, the true infinite is only in the absolute, which is anterior to any composition and is not formed by the addition of parts" (GP V 144).[48] This notion of absolute infinity is non-quantitative in the sense that God or the most perfect being has a non-divisible unity, which admits of no parts; also, it cannot be compared to or measured by any quantity. In this sense, absolute infinity indicated allness and perfection, which cannot be measured. Hence, this notion of infinity is aptly called the Immeasurable or *Immensum*. It involves absolute perfection, completeness as well as unity and indivisibility.

---

[47] Compare with A 6.3 385; LLC 43, where Leibniz articulates the same threefold distinction in slightly different words.

[48] « Le vrai infini à la rigueur n'est. que dans l'absolu, qui est. antérieur à toute composition, et n'est. point formé par l'addition des parties » (GP V 144; NE 2.17.1)

As noted, this conception of the infinite would support Leibniz in avoiding the difficulty facing the notions of infinite number, line, speed, shape, or any other magnitude. Simply stated, on such a conception of infinity, quantitative categories are inapplicable to the essence of true beings (though such things can still be counted).[49] And likewise, maximal quantities cannot be regarded as perfections (see *Discourse on Metaphysics*, paragraph 1). Therefore, if infinity is ascribed to a substance not in a quantitative sense but only in the sense of absolute infinity, the notion of infinite substance or being, qualified in this way, would avoid the contradiction of infinite number and other infinite magnitudes.

While the numerical infinite does not constitute a complete being, infinity in the absolute sense of *Omnia,* does. Leibniz thus reserves the notion of the absolutely infinite for God or the most perfect being.[50] Even though their views about the divine substance are quite different, the connection between infinity and unity is crucial for both Spinoza and Leibniz. In this regard, Leibniz and Spinoza share the following view: substance is the only thing of which one can say that it is infinite, unique, and indivisible. As we have seen, for Descartes, Spinoza, and Leibniz, this conception involves a non-quantitative understanding of the infinity of the most perfect Being.

9.6 Conclusion

Leibniz's encounter with Descartes shows that, while he shares Descartes's view that the only absolutely infinite being is God, he strongly rejects the idea that the infinite must remain inaccessible for the human mind; rather, Leibniz's early rejection of Descartes's distinction between the infinite and the indefinite points to his conviction that we can and ought to investigate the infinite – both in mathematical and metaphysical contexts. Indeed, as we know, Leibniz is using infinity in almost every aspect of his philosophy.

Leibniz's engagement with Galileo' *Discourses on Two Sciences* in 1672–1673 reveals his subtle response to Galileo's paradox. He clearly regards the notion of an infinite number, seen as referring to a whole, as contradictory. At the same time, he is developing a sophisticated way to handle the infinite in mathematics by interpreting both infinitesimals and infinitely large numbers as useful fictions. This analysis also allows him to use infinity in his metaphysics. Leibniz's resolution of Galileo's paradox, however, sets a formidable problem: if the notion of infinite number is inconsistent, doesn't a similar inconsistency affect the notion of an infinite being? This sets Leibniz with a challenge that occupies him in his last years in Paris, namely, to show that the notion of infinite being is possible while that of an infinite number is not. Leibniz's encounter with Spinoza in 1676 (and especially his annotations on Spinoza letter 12) provides some clues about his approach to this challenge.

---

[49] This is at least true in the case of God. The case of created substances is much more delicate. I think that the middle degree is most pertinent in this case. But I cannot elaborate on this here.

[50] See A 6.3 98; LLC 13. For the special status of oneness, see my "Oneness and Substance in Leibniz's Middle Years", *The Leibniz Review*, vol. 24, 2014, pp. 69–94 (with Tamar Levanon).

Leibniz is distinguishing between a quantitative sense of infinity that applies to mathematics and more broadly to whatever can be quantified, and a non-quantitative sense of infinity that applies to metaphysics.

In fact, Leibniz's reading of Spinoza reveals more than this. It reveals that his response to Spinoza (and to his own problem) is more nuanced. As we have seen, he recasts Spinoza's distinction between kinds of infinity, each with a different domain of application, in terms of degrees. As he writes: "I set in order of degree: Omnia; Maximum; Infinitum." Roughly speaking, between the highest degree of infinity, which Leibniz clearly ascribes to the absolute and necessary Being, and the lowest degree of infinity, which he ascribes to quantifiable things, Leibniz invokes a third, intermediate degree of infinity – a maximum of its kind. Leibniz's use of this intermediate degree of infinity must be developed elsewhere.[51]

Abbreviations

A	Leibniz, G. W., *Sämtliche Schriften und Briefe. Hrsg. von der Preußischen* (später: Berlin-Brandenburgischen und Göttinger) Akademie der Wissenschaften zu Berlin. Darmstadt (später: Leipzig, zuletzt: Berlin) 1923 ff. Quote series, volume, page. If not otherwise indicated, the reference is to series 6, volume 3.
AG	Leibniz, G. W., *Philosophical Essays*, eds. and trans. Garber, D., and Ariew, R. (Indianapolis: Hackett, 1989).
AT	Descartes, R., *Œuvres de Descartes*, 11 vols., eds. Adam, C. and Tannery, P. (Paris: J. Vrin, 1996).
Confessio	Leibniz, G. W., *Confessio philosophi, Papers Concerning the Problem of Evil, 1671–1678*, ed. and trans. Sleigh, R. C., JR. (New Haven: Yale University Press, 2005).
CSM	Descartes, R., *The Philosophical Writings of Descartes*, vols. 1,2, eds. and trans. Cottingham, J., Stoothoff, R. and Murdoch., D. (Cambridge: Cambridge University Press, 1984–5).
CSMK	Descartes, R., *The Philosophical Writings of Descartes*, vol. 3, eds. and trans. Cottingham, J., Stoothoff, R., Murdoch., D. and Kenny, A. (Cambridge: Cambridge University Press, 1991).
Curley	*Spinoza, B. de: Collected Works*, vol. 1, ed. and trans. Curley, E. (Princeton: Princeton University Press, 1988).
DesB	*The Leibniz-Des Bosses Correspondence*, ed. and trans. Look, B. C. and Rutherford, D.(New Haven, CT: Yale University Press, 2007).
DSR	Leibniz, G. W., *De Summa Rerum: Metaphysical Papers 1675–1676,* trans. and ed. Parkinson G. H. R. (New Haven and London: Yale University Press, 1992).

[51] I expand on this in my book manuscript *Living Mirrors: Infinity, Unity, and Life in Leibniz's Philosophy*.

E	Spinoza, B. de, *Ethics*.
EN	Galileo, Galilei. *Opera*. Edizione Nazionale, ed. Favaro, A. (Florence, 1898).
Gebhardt	Spinoza B., *Opera*. 4 vols, ed. Gebhardt, C. (Heidelberg: Carl Winter, 1925).
GM	*Die mathematischen Schriften von G. W. Leibniz*, ed. Gerhardt, C. I. (Berlin: Winter, 1860–1875).
GP	*Die Philosophischen Schriften von Leibniz*, 7 vols., ed. Gerhardt, C. I. (Berlin: Weidmann, 1875–90; reprinted Hildesheim: Olms, 1978).
L	Leibniz, G. W., *Philosophical Papers and Letters*, ed. and trans Loemker, L. (Dordrecht: Reidel, 1969).
LLC	Leibniz, G. W., *The Labyrinth of the Continuum. Writings on the Continuum Problem, 1672–1686*, ed. and trans. Arthur, R. (New Haven and London: Yale University Press, 2001).
NE	Leibniz, G. W., *Nouveaux essais sur l'entendement humain*, trans. and eds. Remnant, P. and Bennett, J.(Cambridge: Cambridge University Press, 1981, 2d ed. 1996), cited by book, chapter and section.

References

Adams, R. M. (1994). *Leibniz: Determinist, theist, idealist*. New York: Oxford University Press.

Antognazza, M. R. (2009). *Leibniz: An intellectual biography*. Cambridge: Cambridge University Press.

Ariew, R. (1990). The infinite in Spinoza's philosophy. In E. Curley & P. F. Moreau (Eds.), *Spinoza: Issues and directions; The proceedings of the Chicago Spinoza conference [September 1986]* (Vol. 14, pp. 16–31). Leiden: Brill's Studies in Intellectual History.

Arthur, R. T. W. (2001). Leibniz on infinite number, infinite wholes and the whole world: A reply to Gregory Brown. *Leibniz Review, 11*, 102–116.

Arthur, R. T. W. (2013). Leibniz's syncategorematic infinitesimals, smooth infinitesimal analysis, and second order differentials. *Archive for History of Exact Sciences, 67*, 553–593.

Arthur, R. T. W. (2014). *Leibniz*. Cambridge: Polity Press.

Bassler, O. B. (1998). Leibniz on the indefinite as infinite. *The Review of Metaphysics, 51*, 849–879.

Brown, G. (2005). Leibniz's mathematical argument against a soul of the world. *British Journal for the History of Philosophy, 13*(3), 449–488.

Gueroult, M. (1968). *Spinoza I: Dieu*. Paris: Aubier-Montaigne.

Knobloch, E. (1999). Galileo and Leibniz: Different approaches to infinity. *Archive for the History of the Exact Sciences, 54*, 87–99.

Koyré, A. (1957). *From the closed world to the infinite universe*. Baltimore: John Hopkins Press.

Lærke, M. (2008). *Leibniz lecteur de Spinoza. La genèse d'une opposition complexe, 423–424*. Paris: Honoré Champion.

Levey, S. (1999). Leibniz's constructivism and infinitely folded Matter. In R. Gennaro & C. Huenemann (Eds.), *New Essays on the Rationalists* (pp. 134–162). New York: Oxford University Press.

Nachtomy, O. (2007). *Possibility, agency, and individuality in Leibniz's metaphysics*. Dordrecht: Springer, The New Synthese Historical Library.

Chapter 10
Leibniz's Syncategorematic Actual Infinite

Richard T. W. Arthur

> *"Mea certe philosophia sine infinita actu multitudine stare non potest."—Leibniz to Des Bosses, 14 February 1706; deleted in draft.*

Abstract It is well known that Leibniz advocated the actual infinite, but that he did not admit infinite collections or infinite numbers. But his assimilation of this account to the scholastic notion of the syncategorematic infinite (more accurately, the infinite syncategorematically understood) has given rise to controversy. A common interpretation is that in mathematics Leibniz's syncategorematic infinite is identical with the Aristotelian potential infinite, so that it applies only to ideal entities, and is therefore distinct from the actual infinite that applies to the actual world. Against this, I argue in this paper that Leibniz's actual infinite, understood syncategorematically, applies to any entities that are actually infinite in multitude, whether numbers, actual parts of matter, or monads. It signifies that there are more of them than can be assigned a number, but that there is no infinite number or collection of them (the categorematic infinite), which notion involves a contradiction. Similarly, to say that a magnitude is actually infinitely small in the syncategorematic sense is to say that no matter how small a magnitude one takes, there is a smaller, but there are no actual infinitesimals. In geometry one may calculate with expressions apparently denoting such entities, on the understanding that they are fictions, standing for variable

I am very grateful to David Rabouin for his astute and valuable comments on drafts of this paper, and to Maria Rosa Antognazza for her generous feedback on the penultimate version, which saved me from many errors of misattribution; any remaining errors in the views here expounded are, of course, my responsibility alone.

R. T. W. Arthur (✉)
Philosophy Department, McMaster University, Hamilton, Canada
e-mail: rarthur@mcmaster.ca

O. Nachtomy, R. Winegar (eds.), *Infinity in Early Modern Philosophy*,
The New Synthese Historical Library 76,
https://doi.org/10.1007/978-3-319-94556-9_10

magnitudes that can be made arbitrarily small, so as to produce demonstrations that there is no error in the resulting expressions.

10.1 Introduction

As is well known, Leibniz was a vigorous champion of the actual infinite, insisting on the actual infinitude of monads and of the parts of matter, to name two of the most important instances of the actual infinite in his philosophy. But he denied that any infinite aggregate of terms could constitute a whole or be assigned a number. On his interpretation of the actual infinite, unlike Cantor's, there can be no infinite number, and no infinite sets. In his elaboration of this doctrine, Leibniz appealed to the scholastic distinction between categorematic and syncategorematic understandings of the infinite, denying the infinite in the first sense, and upholding it in the second. He also recognized a third sense of the infinite, the hypercategorematic, which is the kind of all-encompassing infinite that is applicable to God alone.

All of this is well understood, and not a point of contention among the interpreters of Leibniz I shall be considering here. Beyond this bare outline, however, there are some subtle, but in my view crucial, disagreements concerning the interpretation of the syncategorematic infinite. Certain scholars of the calculus have denied that the interpretation of infinitesimals as syncategorematic was Leibniz's mature view, and have seen them as fictions in a different sense. I shall not mainly be concerned with that line of disagreement here, reserving a detailed critique of such views for another occasion. My main concern here is the claim that in mathematics Leibniz's syncategorematic infinite is identical with the Aristotelian potential infinite, so that it applies only to ideal entities, and is therefore distinct from the actual infinite that applies to the actual world. Thus Herbert Breger writes: "Infinity for Leibniz consists only in the (potential) determination that the occasion and the possibility to continue a process always remain. Even if there are infinitely many monads in metaphysics, this does not yet answer the question of the infinite in Leibniz's mathematics"[1]; and much more recently: "There is no actual infinite in Leibniz's mathematics."[2]

In a similar vein Maria Rosa Antognazza has taken Leibniz's syncategorematic infinite, insofar as it "applies to ideal, mathematical entities", to be identical with the potential infinite, so that it is "strictly speaking, only 'indefinite' or 'indeterminate' rather than genuinely infinite" (Antognazza 2015, 10). This kind of infinite, which "concerns the abstract, ideal entities treated by mathematics", is distinguished by her from "the actual infinite which applies to the physical world" (2015, 8). Following the lead of Fabio Bosinelli, Antognazza characterizes the latter actual infinite as syncategorematic in a different sense, where it is understood to apply

[1] "Die Unendlichkeit besteht für Leibniz lediglich in der (potentiellen) Bestimmung, daß der Anlaß und die Möglichkeit zu einer Fortsetzung eines Prozesses stets vorhanden bleiben. Auch wenn in der Metaphysik unendlich viele Monaden existieren, so beantwortet dies noch nicht die Frage nach dem Unendlichen in Leibniz' Mathematik." (Breger 1986, 322).

[2] "Ein Aktul-Unendlich gibt es in der Leibnizschen Mathematik nicht." (Breger 2016a, 124).

distributively ("per modum totius distributivi, non collectivi"), here referencing a famous passage from a note in Leibniz's correspondence with Des Bosses, to which we shall return later.[3] This is a fourth kind of infinity, distinct not only from the potential infinity that is only applicable to ideal entities, but also from the (rejected) categorematic actual infinite, and from the hypercategorematic infinite that applies only to God. But according to Bosinelli this actual infinite that applies to the physical is such that "it cannot be ordered [man kann es nicht ordnen]" (Bosinelli 1991, 168). And according to Antognazza, it "cannot be enumerated, since any enumeration can only be finite. Therefore any enumeration can only yield a traditional syncategorematic potential infinite" (Antognazza 2015, 10). We are thus presented with a dichotomy: on the one side is the infinite in mathematics, which applies only to ideal entities, and is therefore only a potential infinite; and on the other is the infinite pertaining to actual entities, which cannot be ordered or enumerated.

In opposition to these readings, I argue that Leibniz's actual infinite, understood syncategorematically, applies to numbers and to constituents of the actual world in exactly the same way, so that it cannot simply be identified with the Aristotelian potential infinite in mathematics. As I understand Leibniz's meaning, to say that there are infinitely many things in the syncategorematic sense is to say that no matter how many one might take there to be, there are more. In this way, the syncategorematic infinite applies to any entities that are actually infinite in multitude, whether numbers, actual parts of matter, or monads. The word 'actually' qualifies the infinitude of the multiplicity, not the entities that are held to be infinitely many. There is also a corresponding sense in which the infinite and infinitely small in magnitude may be understood syncategorematically, as connoting that no matter how large (resp., small) a magnitude one takes, there is a larger (resp., smaller). This conception licenses working with terms denoting infinitely large and infinitely small magnitudes in geometry; these are fictions whose use is justifiable by demonstrations in which they are replaced by arbitrarily large or small magnitudes. For example, in the calculus a finite magnitude may be taken as a whole with infinitely many infinitely small parts, in the syncategorematic sense: the parts are fictions, defined relatively to the whole of which they are parts, and their use is justifiable by a demonstration that taking sufficiently many sufficiently small parts issues in no error.

In the interests of clarity, I should stress the points I am not contesting. When Breger says that there is no actual infinite in Leibniz's mathematics, he is primarily concerned to deny the reading of the actual infinite in Leibnizian mathematics as categorematic (as in non-Archimedean construals of the continuum and infinitesimals), and I have no quarrel with him (or with Bosinelli or Antognazza) about this. An infinite aggregate of entities does not form a collection or whole, according to Leibniz, so that there is no infinite number assignable to it, no categorematic infinite. I also agree with Bosinelli and Antognazza that the actual infinity that Leibniz sees everywhere in nature is distinct from the categorematic actual infinite and from

[3] "Um eine Identifikation des aktualen Unendlichen mit dem kategorematischen Unendlichen zu vermeiden, muß man es dann" per modum totius distributivi non collectivi "verstehen." (Bosinelli 1991, 165–6).

the hypercategorematic, which latter Leibniz describes as the only "genuine infinite" [vrai infini] (A VI 6, 157). But, I claim, the actual infinite pertaining to the created world, insofar as it is an infinite in multitude, is syncategorematic in exactly the same sense as the actual infinite that applies to numbers, and is distinct from the traditional potential infinite.

An example may help clarify the difference between the syncategorematic infinite and the potential infinite. Take the infinite aggregate of even numbers. On a reading of this infinity as a potential infinity, this means that no matter how many even numbers one assumes there to be, there could always be more. On the reading of the syncategorematic actual infinite that I believe Leibniz is advocating, however, there are actually infinitely many, even though there is no infinite number of them. This is the same concept of actual infinity as applies to the parts into which existing matter is actually divided: more of these parts "can be found than there are unities in any given number" (A VI 4, 1393; LLC 234-235). But if an enumeration of these parts could only be achieved by recourse to the potential infinite, as Antognazza claims, this would threaten to collapse the distinction between potential and actual division, contrary to her intentions.

The case is different when we consider ways in which a number could be divided into fractions, or a continuum into parts. Here we are concerned with possibility, ways in which these entities could be divided, and it is here that the potential infinite applies. The contrast Leibniz offers is between the merely potential, indeterminate, parts of the continuum, and the actual, determinate parts into which existing things are divided (or the infinite unities from which they result), which are discrete. It is a contrast between the continuous and the discrete, not between the mathematical and the non-mathematical. Thus when matter is conceived abstractly (i.e. as a continuum or ideal whole), it is divisible into parts in whatever way one wishes; it is potentially divisible to infinity. But insofar as it actually exists, matter is infinitely divided into a particular progression of actual parts. These actual parts can be ordered and numbered just as the terms of an infinite series could be, so that (contra Breger, Bosinelli and Antognazza) there is not one kind of infinity that applies to mathematical entities, and another that applies to actual things.

This is, no doubt, a subtle difference in interpretation. But I believe it is not without consequences. For as long as Leibniz's syncategorematic infinite is identified with the potential infinite in his mathematics, this will suggest that there is no actual infinite there, and foreclose a very interesting alternative to the Cantorian actual infinite (the transfinite). It will obscure the use Leibniz makes of mathematical analogies in his metaphysics, such as the idea of an infinite series of monadic states. And the idea that any enumeration of the actual infinite must either be finite or require recourse to a potential infinity, will also encourage interpretations of Leibniz's philosophy that deny the actuality of matter for him,[4] and thus sever the link between

[4]One can see this tendency in Bosinelli's characterization of the physical continuum as "ideal because phenomenal": "Leibniz' Unterscheidung zwischen 'continuum physicum' (ideal, weil phänomenal: vgl. Rescher, loc. cit.) und 'continuum mathematicum' (das eine mera possibilitas, ut numeri ist)…" (1991, 165: fn 36).

the division of matter and his argument for the actual infinitude of monads. Each actual part of matter is an aggregate of further actual parts, and, I contend, it is the fact that each such actual part presupposes real unities that establishes that these unities, the monads, are infinitely many. This is in keeping, I argue here, with Leibniz's adherence to the doctrine of the Plurality of Forms, according to which there are already actual parts in a body prior to these parts coming under the dominion of some form, whereas on the Thomist interpretation the parts of a body would be purely potential prior to their receiving a dominant form.

10.2 The Infinite and Infinitely Small as Fictions

Contrary to interpretations that depict Leibniz as having developed his interpretation of infinitesimals as fictions in the 1690s or 1700s in response to criticisms of the calculus,[5] I have argued that he interpreted them this way even as he was formulating the calculus in Paris in 1672–1676, as a natural extension of his work on infinite series.[6] And on my reading, Leibniz's interpretation of infinitesimals as fictions is syncategorematic from that time forth, even if prior to his explicit assimilation of this view to the scholastic tradition. Let me briefly review the main features of this work that are relevant to our discussion.

Leibniz had already been alerted to the contradictions that could be found in treating the infinite as a collection by his reading of Galileo Galilei's *Two New Sciences (Discorsi)* in the Fall of 1672. There, as Leibniz reports, Galileo had demonstrated that there are as many square roots of (natural) numbers as there are natural numbers, and that "therefore there are as many squares as numbers; which is impossible"—an argument now known as Galileo's Paradox.[7] But Galileo had concluded (as had also Gregoire de St. Vincent) "that in the infinite the whole is not greater than the part", an opinion Leibniz says he "cannot accept". The axiom that the whole is greater than its part is, of course, one that Leibniz will hold dear for the rest of his life, and he will always take the fact that infinite wholes contradict this axiom as proving their impossibility.

A particularly clear example of this feature of infinite wholes was provided him later in October 1674, when he was working on Apollonius's hyperbola. By using the method of quadratures—a method that he was in the process of developing into the differential calculus—he was able to show that the difference between two areas under the curve, an infinite one A and a finite one B, come out to be equal to A! For the value of A is expressible as the (divergent) infinite series $1 + {}^1/_2 + {}^1/_3 + {}^1/_4 + {}^1/_5 + \ldots$, an infinite quantity, and B as the (convergent) infinite series $1 - {}^1/_2 + {}^1/_3 - {}^1/_4 + {}^1/_5 + \ldots$,

[5] See for example Jesseph (1998, 2008).

[6] See Arthur (2008) and Levey (2008), both indebted to Ishiguro (1990) and Knobloch (2002). Cf. also Breger: "As early as 1675 Leibniz regarded infinitesimals as fictitious, cf. A VII, 6, 537, 585." (2016a, 157)

[7] Relevant extracts from Galileo's *Discorsi* may be found in (LLC 352-357).

which is finite.[8] But subtracting B—representing an area clearly pictured as a proper part of the area A—from A, yields $1 + {}^1/_2 + {}^1/_3 + {}^1/_4 + {}^1/_5 + \ldots$ again! Leibniz remarks:

> This is rather amazing (*satis mirabilis*), and shows that the sum of the series 1, $^1/_2$, $^1/_3$ etc. is infinite, and consequently that the area of the space [A] remains the same even when the finite space [B] is taken away from it, i.e. that nothing noticeable is taken away.
>
> By this argument it is concluded that the infinite is not a whole, but only a fiction; for otherwise the part would be equal to the whole. (A VII 3, N. 38_{10}, p. 468; October 1674)

Of course, by styling the infinite as a fiction, Leibniz did not mean that it was methodologically dispensable or had no place in mathematics. Indeed, by this time he already knew how to calculate with infinite and infinitely small entities. For example, he had successfully calculated the sum of the very infinite series on which he cut his teeth under Huygens's advice in the Fall of 1672, that of the reciprocal triangular numbers, $T = {}^1/_1 + {}^1/_3 + {}^1/_6 + {}^1/_{10} + {}^1/_{15} + \ldots$. Noting that each of its terms is twice the difference between consecutive terms of the series $R = {}^1/_1 + {}^1/_2 + {}^1/_3 + {}^1/_4 + {}^1/_5 + \ldots$, he used his Difference Principle[9] to calculate the sum T as twice the difference between the first and last terms of the series, $R_1 - R_\infty$, which gives 2. Here R_∞ is the infinitely small infinitieth term, $1/\infty$, and Leibniz has taken $1/\infty = 0$. But he also knew how to justify such calculations rigorously by finding an expression for the yth term, where "*y* signifies any number whatever", calculating with this, and then letting *y* be as large as needed for the error to be smaller than any pre-set quantity. By 1676, Leibniz has formalized this way of dealing with infinite series, and is able to define the sum of a converging infinite series as follows:

> Whenever it is said that a certain infinite series of numbers has a sum, I am of the opinion that all that is being said is that any finite series with the same rule has a sum, and that the error always diminishes as the series increases, so that it becomes as small as we would like. For numbers do not *themselves* [*ipsi … numeri per se*] go absolutely to infinity, since then there would be a greatest number. (A VI 3, 503/LLC 98-99)

Thus even though the terms of such a series cannot be added to form a whole, since there is no last member of the series, one can treat it as if it is an infinite sum provided one can demonstrate that by taking *y* arbitrarily large, the error in truncating the series can be made less than any pre-assigned quantity. Here Leibniz appeals to a principle also appealed to by Newton, that I have termed the Principle of Unassignable Difference (Arthur 2008, 19). It may be expressed as follows:

> Two quantities or ratios of quantities whose difference can be made smaller than any assignable difference (by varying a co-dependent quantity), are equal.

---

[8] Expressing Leibniz's calculation in modern terms, he has calculated an expression for the area A (ACBGM in the original) under the hyperbola $x = 1/(1 - y)$ between the *x*-axis and a straight line parallel to it ($y = 1$) that constitutes one of the curve's asymptotes, by taking the definite integral $\int x\, dy$ between $y = 0$ and $y = 1$ by using an infinite series expansion and integrating term by term. The area B (CBGF on his diagram) is that under the hyperbola between $y = -1$ and $y = 0$, which yields the (convergent) infinite series $1 - {}^1/_2 + {}^1/_3 - {}^1/_4 + {}^1/_5 + \ldots$, which sums to ln2. See Arthur (2013a) for details.

[9] The "Difference Principle" is namely that "the sum of the differences is the difference between the first term and the last" (A VII, 3, 95); see Arthur (2013a, 557).

Thus when the infinite series T, whose yth term is $2/(y^2 + y)$, is truncated after y terms, its sum is $2\{1 - 1/(y + 1)\}$. But by the Archimedean Principle, in place of y we can take a number as large as we please. Application of the Principle of Unassignable Difference then allows Leibniz to say that the sum of the series is 2.

Things are otherwise with the hyperbolic series R, which Leibniz had calculated as representing the area between Apollonius' Hyperbola and its asymptote. The series still has no number of terms, and cannot be understood as a whole. But now it has no finite sum: the Principle of Unassignable Difference cannot be applied. We can say that this (infinite) sum is 1/0, Leibniz writes in his notes on Spinoza's "Letter on the Infinite" in the Spring of 1676, "provided we understand this 0, or nought, or rather a quantity infinitely or unassignably small, instead to be greater or smaller according as we have assumed the last denominator of this infinite series of fractions, which is itself also infinite, smaller or greater" (A VI 3, 282/LLC 114–115). That the infinitieth term is infinitely small is shorthand for saying that it can be made arbitrarily small by taking arbitrarily many terms. The magnitude of the series, however, and of the area it represents, is "greater than we can expound by an assignable ratio to sensible things" (282/115): it has no assignable magnitude.

10.3 The Infinite as Syncategorematic

We have seen that for Leibniz the infinite is not a whole or collection, but one can nevertheless use a term apparently denoting such a thing to say true things concerning the infinite. Elaborating on this idea in later writings, Leibniz noted how this corresponded to certain medieval scholars' conceptions of the infinite as "syncategorematic", that is, as deriving its meaning and its legitimation from the context in which it is used.[10] Thus in the *New Essays* of 1704 he wrote:

> It is perfectly correct to say that there is an infinity of things, i.e. that there are always more of them than can be specified. But it is easy to demonstrate that there is no infinite number, nor any infinite line or other infinite quantity, if these are taken to be genuine wholes. The Scholastics were taking this view, or should have been doing so, when they allowed a syncategorematic infinite, as they called it, and not a categorematic one. (A VI 6, 157/NE 157)

The distinction is grammatical in origin. According to medieval scholars, while certain words (such as nouns) are *categorematic* terms, having meaning in isolation from other words, others (such as conjunctions, quantifiers and articles) are *syncat-*

---

[10] In previous publications I and others had referred to Leibniz regarding the infinite as a "syncategorematic *term*". As Sara L. Uckelman has objected in her thorough treatment of the Scholastics' discussions of the syncategorematic (Uckelman 2015), this is not accurate, as the same term can be used either categorematically or syncategorematically. (I am indebted to João Cortese for bringing this article to my attention.) It is better, she says, to talk of categorematic or syncategorematic *uses* of one and the same term" (2376). It was customary by the seventeenth century, however, to refer to "the syncategorematic infinite" as an abbreviation for "the infinite understood syncategorematically". Leibniz also recognized a third species of the infinite, namely the hypercategorematic infinite—see Antognazza (2015) for a lucid discussion.

egorematic, deriving their meaning from the context in which they are used. Leibniz himself invokes the distinction in his *Matheseos universalis*, Part 1:

> Moreover, just as notions in logic are either categorematic or syncategorematic—e.g. 'man' or 'horse' is a categorematic notion, but in the term 'man and horse' the particle 'and' is syncategorematic—so likewise in Universal Mathesis, to categorematic notions there correspond quantities or numbers which are designated by primary symbols: 1, 2, 3, *a*, *b*, *x*. But to syncategorematic notions there correspond secondary symbols and, so to speak, connotations, like the equals sign or other symbols of relations between quantities. (GM VII 54)[11]

Still other terms, such as *omnis* (all or every), and *infinitum* (infinite), are ambiguous. An expression such as "Infinita sunt finita" is false if *infinita* is taken categorematically ("Things that are infinite are finite"), but true if taken syncategorematically ("Infinitely many things are finite"). Likewise, as Antognazza explains (amplifying on the views of William of Sherwood), *totum* (whole) can be interpreted categorematically, as when we take "*totus mundus est pulcher*" to mean "the world as a whole is beautiful", or syncategorematically, as "when we say, distributively, that 'the whole world (that is, each and every part of the world) is beautiful'." (Antognazza 2015, 7). Although the grammatical distinction between categorematic and syncategorematic uses of a word is more general than that between collective and distributive uses of a quantifier such as *omne*, it is clear that when the adjective "infinite" connotes a whole it is being understood both categorematically and collectively, and that when it is being used syncategorematically, it is also understood distributively.

Turning now to the question of whether the infinite, syncategorematically understood, is to be equated with Aristotle's potential infinite, it is important to realize that there was disagreement in the Schools about how to interpret Aristotle's philosophy of the infinite. Aristotle denied the infinite by addition, but accepted the infinite by division. On the Thomist interpretation of this, as adopted by the Jesuits of Leibniz's time, the latter is simply the potential infinite. This is because matter, taken apart from form, is pure potentiality; matter is only actual or real when it is combined with a form, and it is the substantial form that constitutes the informed matter as one thing. This accords with the axiom that "*being and one are convertible*"; if matter were actually divided, it would be an aggregate of unities, and would no longer be one. So if form unites matter into one being, it can no longer be many; if it were many and not one, it would not *be*, except as a mathematical entity, an abstraction. So matter united by a form can be understood to be infinitely divided provided this is not understood as an actual division; the expression "infinitely divided" can be given a clear sense, but this division cannot be taken in a literal or categorematic sense: it just indicates a potential for being further divided. So thinkers of this persuasion simply equated the syncategorematic with the potential infinite.

But on a competing interpretation of the infinite by division, the infinitude in question is taken syncategorematically and distributively, but as an actual infinite. An influential proponent of such a view was William of Ockham, who claimed that to assert that the continuum has infinitely many parts in a *syncategorematic* sense is

[11] I am indebted to David Rabouin for this reference.

to assert that "there are not so many parts finite in number that there are not more (*partes non tot finitas numero quin plures,* or *non sunt tot quin sint plura)*". As suggested by A. W. Moore (1991, 51), this way of distinguishing the syncategorematic use of the term 'infinite' from its categorematic use can be neatly rendered using modern quantifier logic:

> To say that there are infinitely many things in the *syncategorematic* sense is to say that for any finite number x there is a number of things y greater than this:
> $\forall x \exists y(Fx \rightarrow y > x)$ — x and y range over numbers; $Fx := x$ is finite.
> But there is no *categorematic* infinite, a number of things greater than any finite number: $\exists y \forall x(Fx \rightarrow y > x)$.

That pertains to the infinite in multitude. But there would be a corresponding syncategorematic reading of "infinite" and "infinitely small" when applied to magnitudes, which we could again express with the order of the quantifiers exchanged:

> To say that something is *infinitely large* in the syncategorematic sense is to say that no matter how large it is taken to be (say, of finite magnitude x), it is still larger (of magnitude y with $y > x$): $\forall x \exists y(Fx \rightarrow y > x)$.
> But there are no *categorematically infinite magnitudes*, magnitudes larger than all (finite) magnitudes: $\exists y \forall x(Fx \rightarrow y > x)$.
> To say that something is *infinitely small* in the syncategorematic sense is to say that no matter how small it is taken to be (say, of magnitude x), it is still smaller (of magnitude y with $y < x$): $\forall x \exists y(Fx \rightarrow y > x)$.
> But there are no *categorematic infinitesimals*, magnitudes smaller than all (finite) magnitudes: $\exists y \forall x(Fx \rightarrow y > x)$.

Now, Ockham held that the parts of the continuum (which he conceived as finite) are such that "one must protrude into another" even though they have distinct situations, and that "they must make up a thing that is numerically one, because otherwise they would not be continuous" (*Summa Logica*, Book 1, Chapter 45).[12] So on his view an actually infinite multitude of parts can still constitute one continuum, even though there is no number of all the parts. Since "being and one are convertible", a continuous material whole can therefore have being independently of its being made one by a substantial form, and it can have infinitely many actual parts even in the absence of a substantial form of the whole. As this indicates, Ockham was among those subscribing to the Plurality of Forms. According to these thinkers, in opposition to Thomas, it is possible for the same part of matter to have its own substantial form, and yet be part of the body of another substance. Thus the matter of a corpse can still be actual as a result of the substantial forms of the substances contained in it, even after the soul (the substantial from of the living body) has departed. So on this view, even if in the absence of all forms matter would be pure potentiality, an aggregate of substances that is not unified by a substantial form could still have actual matter—and it could be infinitely divided into actual parts.

[12]—translated by RTWA from text given by The Logic Museum, an online resource accessed April 26, 2017: http://www.logicmuseum.com/wiki/Authors/Ockham/Summa_Logicae/Book_I/Chapter_45). For an account of Ockham on the continuum, see Goddu (1984), esp. p. 208.

One can see some commonality with Ockham's view in Leibniz's early theory of the continuum in his *Theoria Motus Abstracti* of 1672. For there he confidently asserts that "*There are actually parts in the continuum*" and that "*these are actually infinite*" (A VI 1, N41, §1, §2); and this theory also conforms to Ockham's idea that these parts "protrude on one another", since each of two colliding bodies next to one another "endeavours to move into the other's place" so that their endpoints or "beginnings" mutually interpenetrate (§15). Leibniz's innovations were to identify the parts of the continuum with the parts of matter stirred up in the plenum by their differing motions, the "genuine division" acknowledged by Descartes, and also to identify their infinitely small endpoints with Cavalierian indivisibles, following the suggestion of Hobbes.[13] But as we have seen, by 1672 he has come to recognize that there are no indivisibles, if these are taken literally as undivided elements of the continuum, and by 1676 to realize that they must be interpreted syncategorematically as fictional elements of the continuum. But by this time for Leibniz the continuum is purely mathematical: whereas "Matter alone is explicable by a multitude without continuity" (A VI 3, 473/LLC 47), "in the continuum, the whole is prior to its parts; … There is no greatest number, and no least line" (A VI 3, 473/LLC 47).

These statements clearly prefigure Leibniz's mature formulations of how to resolve the problem of the composition of the continuum—for instance, in the formulation he conveyed to Nicolas Remond in July 1714: "In the ideal or the continuum, the whole is prior to the parts … In the real, however, the simple is prior to the aggregates, parts are actual, and are prior to the whole" (GP III 622). His mature position is thus quite distinct from Ockham's, since now the continuum does not have actual parts. Indeed, as has been observed by many commentators, and Breger in particular, Leibniz's views on the continuum are decidedly Aristotelian.[14] The parts of the mathematical continuum are those into which it *can* be divided, but as a continuum it is not actually divided. Leibniz makes this plain (for example) in the elided part of the above quotation from his letter to Remond: "In the ideal or the continuum, the whole is prior to the parts, as the arithmetical unity is prior to the fractions which divide it, and which can be assigned arbitrarily; its parts are merely potential." (GP III 622). The type of infinity characterizing the division of the continuum is thus clearly a potential infinity.

Does this then decide our question, whether Leibniz's syncategorematic infinite is to be equated with Aristotle's potential infinite? If mathematics is the science of magnitude, and such magnitudes are conceived as continua, then one might think that Breger and others must be right to claim that the only operant notion of infinity in Leibniz's mathematics is the potential infinite. I shall argue not, for two reasons.

---

[13] According to Ockham the actually infinitely many parts of the continuum are divisible. Leibniz's wording in the *Theoria Motus Abstracti* is a bit ambiguous about whether the infinitely many parts of the continuum are to be identified with indivisibles and thus compose the continuum; it is possible to read him as holding that the parts are divisible like Ockham's actual parts, and that the indivisibles are only their beginnings.

[14] See in particular Breger's "Le continu chez Leibniz" (1992), reprinted in (Breger 2016a, 127–135).

In the first place, it is *geometry* that is the science of continuous quantity, not arithmetic (and algebra is neutral, since variables may stand for continuous or discrete quantities). This is surely relevant, since Leibniz habitually avails himself of the contrast between the *discrete* quantity pertaining to arithmetic and the *continuous* quantity of geometry to explain his solution to the continuum problem. A case in point is the following passage to De Volder (19 January 1706):

> In actuals there is nothing but discrete quantity, namely the multitude of monads or simple substances, which is greater than any number whatever in any aggregate whatever that is sensible, i.e. corresponds to the phenomena. But continuous quantity is something ideal, which pertains to possibles and to actuals insofar as they are considered possible. For the continuum involves indeterminate parts, whereas in actuals there is nothing indefinite—indeed, in them any division that can be made is made. Actuals are composed as is a number from unities, ideals as is a number from fractions: the parts are actual in the real whole, not in the ideal whole. (LDV 333)

The actual parts are now the discrete parts into which matter is actually divided, as opposed to the ideal parts of the continuum, which are indeterminate, and can be taken as desired. So the mistake underlying the labyrinth of the continuum involves confusing the indeterminate parts of ideal wholes with the determinate parts into which existing things are divided, and looking for the real unities presupposed by actually existing things in what is continuous, and therefore only a mathematical abstraction.

As noted, Bosinelli and Antognazza recognize this actual infinite division of matter, and identify it as an infinite that is syncategorematic and distributive, rather than categorematic. But they understand this to mean that it cannot be ordered (Bosinelli 1991, 168) or enumerated (Antognazza). In Antognazza's words, an enumeration of this distributive actual infinite "can only yield a traditional syncategorematic potential infinite" (2015, 10). Bosinelli, for his part, bases his analysis on the distinction between the "*corpus mathematicum*" of the Cartesians, criticized by Leibniz, and the "*corpus physicum*" he endorsed. The unities to be found in the former are "*unitates irreales*"; they are the fictional unities of the calculus, the infinitesimal line segments dx which integrate to form a finite line segment x, of which they are homogeneous parts. This contrasts with the "*unitates reales*" in physical bodies, the monads, which, lacking extension, "have no geometrical dimension, and which are the unities of the 'heterogenea'". This is a nice contrast; but it fails to acknowledge that in neither case are the unities the result of a division. One does not perform a potentially infinite division to generate infinitesimals, the "*unitates irreales*": the homogeneous parts of a finite geometrical body are still finite, even if arbitrarily small. Rather one posits infinitesimals as fictional infinitely small elements; they are not homogeneous parts of the continuum, since they have no ratio to the continuum of which they are elements, and are therefore homogonous to it. And as for the "*unitates reales*", one does not perform an actually infinite division to generate monads, since, as Bosinelli of course acknowledges, they are not component parts of bodies, but only constitutive unities thereof.[15]

So what, then, are the determinate parts in an actually infinite division? Leibniz is clear: they are the parts into which existing bodies are divided by their differing motions. He reiterates this on numerous occasions in both published and unpublished papers. "Every portion of matter is not only divisible to infinity, as the ancients realized, but is again actually subdivided without end, each part into further parts, each one of which has its own motion" (*Monadology*, §65; GP VI 618). But in the following unpublished fragment the syncategorematic nature of the actually infinite division is particularly evident:

> Created things are actually infinite. For any body whatever is actually divided into several parts, since any body whatever is acted upon by other bodies. And any part whatever of a body is a body by the very definition of body. So bodies are actually infinite, i.e. more bodies can be found than there are unities in any given number. (c. 1678-9; A VI 4, 1393; LLC 235)

This is exactly the same notion of the actual infinite in multitude as occurs in his mathematics: the number of parts (or bodies) is actually infinite in the sense that for any finite number, there are actually (not merely potentially) more bodies than this. It is also precisely the wording Leibniz had used in the *New Essays* to describe what the Scholastics meant, or "should have" meant "when they allowed a syncategorematic infinite": "an infinity of things, i.e. that there are always more of them than can be specified". I take the implicit criticism in the "should have meant" to be that many Scholastics had followed Aristotle in identifying the *syncategorematic infinite* with the *potential infinite*, with the result that they denied the actual infinite.

As noted above, Bosinelli claimed that this kind of infinity "cannot be ordered". But the parts into which actually existing matter is divided by the differing endeavours of each part is a determinate partition, one among all possible partitions of that matter. Existing matter, Leibniz explains in the 1689 tract known as "First Truths", "cannot be divided in all possible ways, because not all creatures are in the same part, but only a certain infinite progression of them. In the same way, if you were to bisect a straight line and any part of it, you would set up different divisions than if you were to trisect it" (A VI 4, 1648/L 270). So when Leibniz tells De Volder that "every division that can be made, is made" he means that there is a specific infinite partition of determinate parts at each distinct instant. A "certain infinite progression" of parts is ordered in the same way as an infinite series is, and in principle there is a law governing any such progression. Consequently, the parts of such an infinite progression could in principle be numbered, just as can the terms of an infinite series—S_1, S_2, S_3, etc.—however haphazard the law of the series may be. When Leibniz tells Des Bosses that an infinite aggregate of such parts "cannot be enumerated" (LDB 32–33)—or "has no number" [*neque constare numero*]—he means that

[15] "Deshalb sind Monaden die Einheiten des „corpus physicum", aber sie sind nicht etwa seine Komponenten, denn es gibt keine komponierte Substanz, die eine „metaphysische Einheit"eine eigentliche „realitas"hat." (Bosinelli 1991, 153).

there is no overall number of constituents in the aggregate, as there would be if it were a whole or a collection, not that the parts could not in principle be assigned numbers severally.[16] Thus even though there is no number of all the parts, just as there is no number of terms in an infinite series, there are actually infinitely many constituents of each, in the syncategorematic sense. Thus as Hidé Ishiguro has affirmed, for Leibniz "there are infinitely many substances and infinitely many numbers in his sense of infinitely many, i.e., actually more than any finite number of them, and not merely potentially more" (Ishiguro 1990, 80).

Breger and Antognazza both object that since numbers are only ideal entities, they cannot be actually infinite. Writes Breger: "If numbers are ideal possibilities, however, it is difficult to see in what sense one could speak of the actual infinite in Leibnizian mathematics."[17] Similarly, commenting on Leibniz's use of the example of the multitude of even numbers to illustrate the actual infinite, Antognazza interprets Leibniz as "offering a mathematical *analogy*, as opposed to maintaining that the *actual* infinite (even if thought of syncategorematically) applies to mathematical, abstract entities and to the ideal, mathematical continuum" (2015, 9). But, again, these objections fail to distinguish the discrete and countable entities of arithmetic, such as integers, from the indeterminate parts of continuous entities. Leibniz himself is explicit that the same notion of actually infinite multitude applies to the natural numbers as to actually existing things. It is not just that there *could* always be more numbers, but there actually *are* infinitely many, existing as real possibilities in the divine mind. "For it cannot be denied that there are really natures of all possible numbers, at least in the divine mind, and thus that the multitude of numbers is infinite" (LDB 32/33).

So numbers are actually infinite. But, I contend, this does not mean that they have actual existence. When Leibniz says there are "actually infinitely many *x*", I take the "actually" to modify the "infinity", not the *x*. The ontological status of the *x* is a separate matter. Numbers, like relations and truths, exist in the divine mind, but as possibles, not as actual existents. Leibniz ascribes reality to relations taken in abstraction, including orders such as space and time, but this reality consists in existence in the divine mind, as opposed to existence in the created world. Thus insofar as "actual" qualifies existence, I take it to connote what has been created, as opposed to remaining merely possible. The actual is not restricted to monads: Leibniz frequently talks of actual phenomena, as opposed to merely possible phenomena, and actual divisions of matter, as opposed to merely possible ones.[18]

---

[16] Maria Rosa Antognazza assures me that she agrees with me here (private communication); but insists that the *actual* infinite cannot apply to *ideal* entities like numbers (see below for further discussion of that objection).

[17] "Wenn Zahlen aber ideale Möglichkeiten sind, dann ist schwer zu sehen, in welchem Sinne man vom Aktual-Unendlichen in der Leibnischen Mathematik sprechen könnte." (Breger 1986, 322)

[18] Also, not all monads are actual. For monads are the basic ontological units of the created world—but also of any possible world. Monads in a possible world are only actual counterfactually: they would be actual, were that world to be the one that God deemed the best and created. So monads are neither the only actuals, nor are they all actual.

10.4 Syncategorematic Infinitesimals

My second reason for objecting to the identification of the syncategorematic infinite in Leibniz's mathematics with the potential infinite is that the latter does not exhaust the meaning of the former even in geometry. As we have seen, notwithstanding Leibniz's denial that there are any infinite or infinitely small mathematical quantities such as areas or lines "if these are taken to be genuine wholes", he was able to calculate with an infinity of infinitely small parts to derive results about finite lines and areas. In such calculations, however, the supposition of infinitely small lines has no meaning outside of their being supposed in such a construction, where an infinity of them is fictionally supposed to compose to a finite quantity. As Hidé Ishiguro was I believe the first to point out, the syncategorematic interpretation of a mathematical quantity as consisting in an infinite multitude of infinitely small parts must therefore be balanced by a corresponding syncategorematic reading of these infinitely small parts.[19]

Now geometric infinitesimals are syncategorematic in the sense that they are defined contextually, deriving their meaning from the relations in which they stand. In this sense they are analogous to other mathematical fictions, such as imaginary roots. They are *compendia loquendi*, abbreviations of thought, allowing various expressions to be derived more easily and economically. Terms apparently referring to categorically infinitely small entities actually do not so refer. Instead, when the context for their successful use is explicated, they are interpretable as standing for relations among finite magnitudes. Still, unlike imaginary roots, the sense in which infinitesimals are fictions for Leibniz is elaborated in terms of the syncategorematic infinite outlined above. There are no actually infinitely small parts in the categorematic sense: such a notion is contradictory. But one can still work with terms that apparently refer to such entities as fictions; and it is the syncategorematic understanding of the actually infinitely small—as a part so small that a smaller can always be found—that licenses their use. To justify using them, "it suffices to substitute for the infinitely small something as small as one wishes, so that the error is smaller than any given" (GP II 305/LDB 32-3). They are not themselves substituted by finite quantities while we are using them as fictions; but the legitimacy of so using them as *compendia loquendi* can be justified by such substitution, although only in the context in which they occur.

The fact that infinitesimals are *parts only in relation to the continuous whole* of which they are fictionally supposed to be parts explains why Leibniz, in trying to justify their use, resorted to the notion of *incomparables*.[20] He introduces this term

---

[19] In the second edition of her *Leibniz's Philosophy of Logic and Language*, Ishiguro claims that Leibniz "maintained that one can have a rigorous language of infinity and infinitesimal while at the same time considering these expressions as syncategorematic (in the sense of the Scholastics), i.e., regarding the words as not designating entities but as being well defined in the proposition in which they occur" (1990, 82), citing Leibniz's words to this effect in his letter to Varignon of 1702 (GM IV 93).

[20] Leibniz's talk of "incomparably small magnitudes" has drawn the following rebuke from Ishiguro: "It is misleading for Leibniz to call these magnitudes [such as the grain of sand in relation

in his essay *Tentamen de motuum coelestium causis* published in the *Acta* of Leipzig in 1689. There he justifies his use of differentials to derive the inverse square law of gravitation in a passage that he later refers to as his "*Lemmas on Incomparables*" as follows:

> In the demonstrations I have assumed *incomparably small quantities*, such as the difference between two common quantities, incomparable with the quantities themselves. For, if I am not mistaken, such matters can be expounded most lucidly in this way. Thus if someone does not want to employ *infinitely small* quantities, one can take quantities as small as one judges sufficient for them to be incomparable, so that they produce an error of no importance, or rather one smaller than a given error. (*Tentamen*, 1689; GM VI 150-1, 168)

For example, using the characteristic triangle for a given curve whose ordinate y is some function of its abscissa x, say $y = 2x^2/a - x$, the slope of the tangent is given by $dy/dx = 4x/a - 1 + 2dx/a$. Here the last term on the right is incomparable with $4x/a - 1$, and can therefore be set to 0. dy and dx are unassignables, incomparably small in relation to y and to x, but having a finite ratio to one another of $(4x - a)/a$, equal to the slope of the tangent at the point (x, y).

Thus one can proceed methodologically as if there are infinitely small quantities, on the understanding that a syncategorematic justification can always be given. Or, one can insist on dispensing with them and working with variable finite quantities alone, using the Archimedean axiom and the Principle of Unassignable Difference to complete the demonstration. This parallels the modern attitude to the use of differentials (as in group theory, or in differential geometry): one calculates with them as if there are such entities, on the understanding that the calculations can be justified using ε-δ techniques; or, if one is a purist, one can proceed with finite quantities alone, using the theory of limits.

In this understanding of Leibniz's infinitesimals as incomparables, I believe I have no disagreement with either Breger or Bosinelli. We all agree that an interpretation of them as incomparables means that they are fictions. Insofar as they are defined only in relation to other terms, and are substitutable by variable quantities rather than fixed elements of the continuum, they are incompatible with the non-Archimedean theories of the infinitesimal and the continuum proposed by Abraham Robinson and more recent authors.[21] It is the Archimedean property of geometric magnitudes, the fact that geometric quantities may be taken as small or as large as one pleases, that provides the foundation for Leibniz's Law of Continuity. And this is clearly consonant with the Aristotelian character of Leibniz's continuum, whose

to the globe of the Earth that he mentions to Varignon (GM IV 92)] incomparably small. What his explanation gives us is rather that a certain truth about *comparably* smaller magnitudes gives rise to the notion of incomparable magnitudes, not incomparably smaller magnitudes. If magnitudes are incomparable, they can be neither bigger nor smaller." (Ishiguro 1990, 87–88). This may be true of the grain of sand and the Earth; but what Leibniz means by his "incomparable" is that a magnitude like dx is (demonstrably) of no assignable magnitude in comparison with x: the relation between them (which involves comparison) is unassignable, i.e. smaller than any that can be assigned. See Breger (2016b) for an illuminating discussion of his matter.

[21] See Herbert Breger, "Leibniz's Calculation with Compendia", reprinted in Breger (2016a, 147–158); and Bosinelli (1991, 156–159); and especially Rabouin (2015).

parts are "indeterminate", "can be assigned arbitrarily", and are "merely potential".

Where we differ, again, is on the interpretation of the term 'syncategorematic'. According to Bosinelli, Leibniz "ascribes to the 'infinitesima' only a 'syncategorematic infinite', which he identifies with the potential infinite".[22] But the syncategorematic infinitesimal is not merely an indeterminate finite part: it is a fictional actually infinitely small element, whose use may be justified by exploiting the Archimedean property of the continuum, that is, by substituting for it sufficiently many sufficiently small parts to demonstrate that the error is null.

Explaining this to Pierre Varignon in February 2, 1702 Leibniz says that it had not been his design "to assert that there are in nature infinitely small lines in all rigour, or compared with ours, nor that there are lines infinitely greater than ours". This is why, in order to avoid these subtleties, he had proposed that "it suffices to explain the infinite by the incomparable, that is to say, to conceive quantities incomparably greater or smaller than ours" (GM IV 91/L 543). Nevertheless, it should not be thought that this reduces the science of the infinite to fictions,

> for there always remains a syncategorematic infinite, as they call it in the Schools, and it remains true that 2 is as much as $^1/_1 + ^1/_2 + ^1/_4 + ^1/_8 + ^1/_{16} + ^1/_{32} + \ldots$, which is an infinite series in which all the fractions whose numerators are 1 and whose denominators are in geometric progression of powers of 2 are comprised at once, even though only ordinary numbers are used and no infinitely small fraction, or one whose denominator is an infinite number, ever occurs in it. (93/544)[23]

As we have seen, this is precisely the understanding of the sum of an infinite series that Leibniz had reached already in 1676; and here he explicitly confirms that he regards it as syncategorematic. There is an actual infinity of terms, otherwise the sum would not reach 2, but would be some finite amount short of 2. So it is not correct to say that this syncategorematic infinite is a potential infinite "since any enumeration could only be finite" (Antognazza 2015, 10), for there are infinitely many terms; and this is the case despite the fact that there is no number of all the terms in the series. Nor is there is an actually infinitely small last term, even though the sum can be computed (using the Difference Principle) by supposing a fictionally infinitely small last term. And this assumption may be justified by taking a quantity sufficiently small as to produce no error (using the Principle of Unassignable Difference). As Leibniz wrote in the *Mémoires de Trévoux*:

> For in place of the infinite or infinitely small one takes quantities as great and as small as necessary for the error to be less than the given error, in such a way that it differs from the

---

[22] "Er schreibt den 'infinitesima' nur ein 'synkategorematisches Unendliches' zu (GM IV, 92-93), das er mit dem potentiellen Unendlichen identifiziert…" (Bosinelli 1991, 157).

[23] "il ne faut point s'imaginer que la science de l'infini est. degradée par cette explication ct reduite à des fictions; car il reste tousjours un infini syncategorematique, comme parle l'ecole, et il demeure vray par exemple que 2 est. autant que $^1/_1 + ^1/_2 + ^1/_4 + ^1/_8 + ^1/_{16} + ^1/_{32} + \ldots$, etc. ce qui est. une serie infinie, dans laquelle toutes les fractions dont les numerateurs sont 1 et les denominateurs de progression Geometrique double, sont comprises à la fois, quoyqu'on n'y employe tousjours que des nombres ordinaires et quoyqu'on n'y fasse point entrer aucune fraction infiniment petite, ou dont le denominateur soit un nombre infini." (GM iv 93/L 544)

style of Archimedes only in the expressions, which are more direct in our method and more conformable to the art of discovery.[24]

The above interpretation of infinitesimals and the syncategorematic actual infinite is of the greatest relevance to the applicability of the calculus to physical reality. Bosinelli leaves us with only a vague idea of how the unreal unities of the calculus are supposed to be related to the real unities in physical bodies. He bypasses the actually infinite parts of matter in comparing the *corpus physicum* (a monadic aggregate, understood distributively) with the *corpus mathematicum* (an infinite aggregate of infinitesimals, understood as indefinite) (Bosinelli 1991, 168). But it is precisely because infinite division is understood syncategorematically as issuing in parts that are always further subdivided, that heterogeneous actuality can be accurately represented by parts that can be made as small as desired so as to produce no resulting error. Thus a curved trajectory is in fact discontinuous, since there is no uniform rectilinear motion that is not interrupted. Consequently it can be represented as a sequence of contiguous arbitrarily small rectilinear uniform motions that are sufficiently small so as to produce no error. Likewise, a monadic series of perceptions consists in successive finite perceptions, each apparently uniform, although divided within by changes of which the perceiver is unaware, "because these impressions are either too small and too numerous, or too unvarying" (A VI 6, 53); it is because this division proceeds internally without limit that the series does not violate the law of continuity. Thus it is the syncategorematically actually infinite division of the phenomena of nature that accounts for their susceptibility to a description by mathematics, as Leibniz explains to Bayle: "the actual phenomena of nature are arranged, and must be, in such a way that nothing ever happens that violates the law of continuity … or any other of the exact rules of mathematics" (GP IV 568-9).[25]

10.5 The Fourth Infinite: Leibniz and Des Bosses

In Sect. 10.3 above I argued that Leibniz abandoned his early conception of the continuum as containing actually infinitely many actual parts in favour of an Aristotelian conception of the continuum as containing merely potential parts. This would be in keeping with a Thomist conception of the infinite division of matter as indicating merely a potential for being further divided. Is it possible, then, that in abandoning his first theory of the continuum, Leibniz turned his back on the plurality of forms and reverted to a Thomist view, where the body of a substance can be

---

[24] "Car au lieu de l'infini ou de l'infiniment petit, on prend des quantités aussi grandes et aussi petites qu'il faut pour que l'erreur soit moindre que l'erreur donnée, de sorte qu'on ne diffère du stile d'Archimède que dans les expressions, qui sont plus directes dans nôtre méthode et plus conformes à l'art d'inventer." (GM V 350)

[25] For an excellent account of the relationship of Leibniz's geometry to the phenomena, see Douglas Marshall's (2011).

understood as infinitely divided only in a mathematical sense, the division being purely potential? We can find explicit answers to these questions if we turn to a consideration of Leibniz's correspondence with the Jesuit Father, Bartholomew Des Bosses.

In fact, it is in this correspondence that we find the main textual basis for the identification of the syncategorematic infinite with the potential infinite in mathematics by Bosinelli and Antognazza. The following passage is from a "Supplementary Study"[26] included with his draft of a letter to Des Bosses of 1 September 1706, but crossed out and not sent with the letter:

> *There is a Syncategorematic infinite*, that is, a passive power having parts, namely the possibility of further progress in dividing, multiplying, subtracting, or adding. *There is* also *a Hypercategorematic infinite*, or potestative infinite, an active power having parts, as it were, eminently but not formally or actually. This infinite is God himself. But *there is not a Categorematic infinite*, that is, one having actually infinite parts formally.
>
> There is also an actual infinite in the sense of a distributive whole, not a collective one. Thus, something can be stated of all numbers, though not collectively. In this way it can be said that for every even number there is a corresponding odd number, and vice versa; but it is not therefore accurately said that there is an equal multitude of even and odd numbers. (LDB 52-3/GP II 314-15)

Here Leibniz begins by identifying the syncategorematic infinite as potential, but then introduces his interpretation of an actually infinite multitude as syncategorematic in the sense that it must be understood distributively, not as a collection or a whole. Bosinelli and Antognazza take this to be a distinct species of the infinite that applies only to the actual, in distinction from the syncategorematic infinite that applies only to ideal entities, like numbers.

There is already a problem with this reading in that the example Leibniz gives to illustrate the actual infinite in the distributive sense precisely concerns numbers. To repeat: "In this way it can be said that for every even number there is a corresponding odd number, and vice versa; but it is not therefore accurately said that there is an equal multitude of even and odd numbers." (53/315) This paragraph thus appears to be a gloss on what it means to say that there are actually infinitely many entities of any kind. When we understand that claim syncategorematically, we are not referring to a collection, and so are necessarily using the term distributively—this is what we saw above in the example given by Antognazza herself in explaining the position of William of Sherwood. What I take Leibniz to be saying in the above paragraph, therefore, is that there are actually infinitely many even numbers in exactly the same sense as there are actually infinitely many parts of matter: however many you take there to be, there are more. Antognazza acknowledges the awkwardness for her reading of Leibniz's using numbers in an example of an actual infinite that is not supposed to apply to numbers, casting it as a mere analogy, as we saw above. But rather than dwell on that, let's turn to the Leibniz-Des Bosses correspondence to set the above passage in context.

[26] See the editors' note 8, LDB 409.

At the beginning of the correspondence, Des Bosses lays his cards upon the table. The following, he claims, are the five axioms "on which the Aristotelian philosophy depends"—axioms taken from Aristotle's metaphysics which, he tells Leibniz in his letter of 12 February 1706, "I believe agree well enough with your system":

> 1. Being and one are convertible. 2. The continuum is infinitely divisible. 3. There is no actual infinity in nature. 4. Unity is the principle of number. 5. In causes and principles there is no progression to infinity but always some stopping point. (LDB 14-15).

Leibniz is unequivocally opposed to axiom 3. As he tells Des Bosses in his reply of 14 February: "That there is an actual infinity in nature I do not doubt; and assuming the plenitude of the world and equable divisibility of matter, it follows from the laws of varying motion that any point whatever is moved by a motion different from any other assignable point." (As we saw above, this was Leibniz's argument for the actually infinite division of matter in the syncategorematic sense.) He was going to add "My philosophy certainly cannot stand without an actually infinite multitude" (the quotation at the head of my paper), but thought better of it, opting for the less confrontational "Nor do I see why we should try to avoid this conclusion" (LDB 20-21).

But he probably had some inkling of why Des Bosses thought one should avoid it. For in his letter of 12 February Des Bosses had tied his (or, perhaps, only his Order's) Thomist colours to the mast, urging Leibniz to agree with St. Thomas's teaching that "in man there may be many forms according to essence but only one according to existence" and that "it is impossible for two things that are without qualification substantial to exist in the same material composite" (LDB 18-19). But this would be incompatible with Leibniz's understanding of the infinite division of any body (including that of a substance) as issuing in actually infinitely many determinate parts, each presupposing a principle of unity. For then, given the convertibility of being and one, there would be many unities in every body, not just potentially, but in actual existence. So Leibniz appends a postscript to his reply, reaffirming his position:

> Since monads or principles of substantial unity are everywhere in matter, it follows from this that there is also an actual infinity, for there is no part, or part of a part, that does not contain monads. (LDB 14-15).

Here we see precisely why Leibniz's philosophy "cannot stand without an actually infinite multitude". If the division of matter in a substance were such that it contained only potential parts, there would be no reason to say that it had anything other than a finite number of parts; and there would be no argument for the infinitude of monads.[27]

[27] This is precisely the line I take in Arthur (2011), where I argue that the actually infinite division of body is the basis for Leibniz's claim that there are in any body actually infinitely many monads, where the infinite is understood syncategorematically.

In his reply of 2 March, recognizing Leibniz's adherence to the actual infinite in multitude, Des Bosses trots out the Thomist argument that Leibniz had perhaps anticipated:

> To begin: if being and one are convertible, then nothing will exist simply and actually in reality except what is actually and simply one. But a fraction of unity or of what is actually and simply one, is not actually and simply one…
>
> Again: any part of matter exists, therefore any part of matter is one or many; if many, a part of a part is one, for where there is not one, there are not many. Moreover, that which is one is not many. Therefore, matter, insofar as it remains with one entelechy, is not actually many. (LDB 26–27).

Leibniz had already tried to head off this line of argument in his previous letter by insisting that although he agrees that "being and one are convertible, and that unity is the principle of number", this is only so "if you consider reasons, or priority of nature, and not if you consider magnitude: for we have fractions, which are certainly less than unity, into infinity" (LDB 20-21). In his reply to the above argument (11 March) he insists that "Numbers, unities and fractions have the nature of relations. And to that extent they can in some way be called 'beings'."[28] In the same way, a material unity may be considered to be composed of fractions, "as the value of one groschen is an aggregate of the values of two half-groschen" (LDB 30-31). But this is not so of substances:

> A fraction of an animal, or a half-animal, therefore, is not one being per se, since this can be understood only of the body of an animal, which is not one being per se but an aggregate, and has an arithmetical unity and not a metaphysical unity. But just as matter itself, if it lacks an entelechy adequate to it, does not make one being, so neither does part of it. Nor do I see what would prevent many things from actually being subject to one entelechy; on the contrary, that is necessarily so. (LDB 30-31)

So Leibniz explicitly rejects the Thomist interpretation of the nature of matter, where an entelechy or form bestows actuality on an otherwise merely potential and undivided matter, and opts for the plurality of forms view. And this is because (secondary) matter is actually divided into parts, and not merely potentially divided, and each of these parts contains entelechies (monads):

> When I say that there is no part of matter that does not contain monads, I illustrate this with the example of the human body or that of some other animal, any of whose solid and fluid parts contain in themselves in turn other animals and plants. And this, I believe, must be said again of any part of these living beings, and so on to infinity. … For matter changes like a river, with the entelechy persisting as long as the machine persists. The machine has an entelechy adequate to it, and this machine contains other machines obviously inadequate to the primary entelechy, but nevertheless endowed with their own entelechies adequate to them, and separable from the prior whole. (LDB 34-35)

In the same letter of March 2 Des Bosses had tried another tack, asking whether Leibniz thought it necessary "to admit in nature an actual infinite in magnitude in

---

[28] For a lucid treatment of Leibniz's concept of number, see Kyle Sereda (2015). Given that numbers are relations, he argues, they have being in exactly the same sense as relations: they are "inhabitants of the divine mind, have reality independent[ly] of the created world, express a certain sort of possibility, and are the subject of necessary truths" (Sereda 2015, 46).

exactly the same way as in multitude" (to Leibniz, 2 March 1706; LDB 26/27). Remarking that their mutual acquaintance Giovanni Battista Tolomei had rejected the former while defending the latter, he asks whether Leibniz agrees. He notes Leibniz's claim in the "*Mémoires de Trévoux*" that "it is not necessary that the infinite be taken rigorously",[29] and quotes back to Leibniz his comment in his *Specimen dynamicum* about his use of differentials to represent dead force. There Leibniz had written: "... although I would not claim on this account that these mathematical entities are such as are actually found in nature, but only that they are useful for making accurate estimates by means of a mental abstraction" (GM vi 238/LDB 26/27). Taking these two pronouncements together, Des Bosses wrote, "I would have conjectured that the infinite that you build upon is confined to the syncategorematic, for what prevents us from transferring what you say about degrees of impetus to a multitude of substances?" (LDB 26/27). In other words, Des Bosses is suggesting, perhaps Leibniz intends his actual infinite in multitude to indicate only an indefiniteness of further addition or division.

We have already seen that it cannot be so interpreted. For Leibniz existing matter is actually divided into a syncategorematic infinity of parts, and not merely potentially divided into parts like a mathematical body. Each of these parts is actual and discrete, and each such part contains entelechies (monads). But that pertains to the *multiplicity* of parts. What about Des Bosses's question about the infinite in magnitude?

In reply Leibniz makes a distinction: "An actual infinity in magnitude cannot be shown in the same way as it can in multitude." (LDB 30/31). Given the understanding of the actual infinite I have outlined above, we can see why. On that interpretation of the actual infinite as syncategorematic, it is unproblematic to say that the world is actually infinite in the sense of containing more (embodied) monads than can be specified. But this does not constitute an infinite whole, any more than does the area under Apollonius's hyperbola, and so no magnitude can be assigned to it. Thus Leibniz writes to Des Bosses:

> Arguments against an actual infinity assume that if this is admitted, there will be an infinite number .... But it must be recognized that an infinite aggregate is not one whole or endowed with magnitude, and has no number [*neque constare numero*]. And accurately speaking, instead of 'infinite number' it should be said that there are more than can be expressed by any number, and in place of 'infinite straight line', that it is a line produced beyond any magnitude that can be assigned, so that there is always a longer and longer straight line. It is of the essence of number, of line and of any whole whatsoever, to be bounded. Consequently, even if the world were infinite in magnitude, it would not be one whole... (to Des Bosses, 2 March 1706; GP II 305; LDB 30-33)

Leibniz cedes that there can be an actually infinite in magnitude, provided this is understood syncategorematically, i.e. not as a greatest magnitude, but as one so

[29] In his "Mémoire de Mr. G. G. Leibnitz touchant son sentiment sur le calcul différential" (*Mémoires de Trévoux*, November 1701; GM V 350), Leibniz had written "J'ajouterai même à ce que cet illustre Mathématicien [sc. L'Hospital] en a dit, qu'on n'a pas besoin de prendre l'infini ici à la rigueur, mais seulement comme lorsqu'on dit dans l'optique, que les rayons du Soleil viennent d'un point infiniment éloigné, et ainsi sont estimés parallèles."

great as to be greater than any finite magnitude that might be assigned. But this would still not constitute one whole, "nor could it be imagined, as certain ancient authors did, that God is the soul of the world, not only because he is the cause of the world, but also because such a world would not be one body" (305/32-33). Turning next to Des Bosses's remarks about his calculus, Leibniz replies

> Speaking philosophically, I maintain that there are no more infinitely small magnitudes than there are infinitely large ones, that is, no more infinitesimals than infinituples. For I hold both to be fictions of the mind through an abbreviated way of speaking, adapted to calculation, as imaginary roots in algebra are too. Meanwhile I have demonstrated that these expressions are very useful for abbreviating thought and thus for discovery, and cannot lead to error, since it suffices to substitute for the infinitely small something as small as one wishes, so that the error is smaller than any given, whence it follows that there can be no error. (GP II 305/LDB 32-3)

Here we see the denial of the actually infinitely small clearly explained in terms of what I have called the Principle of Unassignable Difference. One can use an expression involving a (fictionally) infinitely small quantity to abbreviate thought and facilitate discovery on the understanding that one could substitute for it something that is so small that the resulting error can be made smaller than any given error. This is the syncategorematic infinitesimal described above.

Nevertheless, as Antognazza points out, the parts of matter are *actual*, according to Leibniz, whereas the parts of a continuum are regarded by him as merely *ideal*, as possible parts into which it could be divided. As we have seen, this is the main motivation (on her part as well as Bosinelli's and Breger's) for insisting that the syncategorematic infinite that applies to continua and the ideal entities of mathematics is distinct from the syncategorematic actual infinite that applies to actual entities. In support she quotes the continuation of the passage in Leibniz's letter to Des Bosses, which passes from a discussion of mathematics to "the realities of physics":

> To pass now from the ideas of geometry to the realities of physics, I hold that matter is actually fragmented into parts smaller than any given, that is to say, that there is no part that is not actually subdivided into other parts undergoing different motions. This is demanded by the nature of matter and motion, and by the whole frame of the universe, for physical, mathematical and metaphysical reasons. (to Des Bosses, 2 March 1706; LDB 32-35)

Here, it is true, a distinction needs to be made. But it is that between the infinitely small parts of matter resulting from an actual infinite division, understood syncategorematically, and the fictional infinitesimals of his mathematics. Thus concerning the realities of physics, a "part smaller than any given" is interpreted syncategorematically as connoting that there is "no part that is not actually subdivided into other parts", rather than that there is an infinitely small magnitude in a categorematic sense; whereas the infinitely small magnitudes of geometry are neither actual parts nor potential parts, but "fictions of the mind through an abbreviated way of speaking", for which one is able to substitute in expressions containing them "something as small as one wishes, so that the error is smaller than any given", in keeping with the syncategorematic interpretation of infinitesimals.

10.6 Conclusion

I have argued that the syncategorematic infinite in Leibniz is not to be equated with the potential infinite of traditional Aristotelian metaphysics. While the potential infinite is clearly appropriate to the division of the continuum, where the parts are indeterminate, Leibniz understands the syncategorematic infinite in such a way that it is applicable to any entities that are actually infinite in multitude. The actually infinite division of actual matter, syncategorematically understood, issues in infinitely small parts that are not so small that there are not smaller (finite) parts. Each such part is determinate and actual, and presupposes a principle of substantial unity or monad, thus underwriting the syncategorematic infinitude of monads. In the calculus, an infinitesimal is a fictionally incomparably small element of the continuum, whose use in calculations is justified by the Archimedean property of magnitudes. Thus there is an actual infinite, syncategorematically understood, distinct from either the categorematic or the hypercategorematic infinites. And there is also a syncategorematic infinitesimal in mathematics, although this is not a potential part of the continuum but a fictional infinitely small part whose application to the natural world is justifiable on the syncategorematic interpretation.

This by no means exhausts the points of interest about Leibniz's theory of the infinite. Aside from the hypercategorematic infinite, there are many questions I have not treated here concerning the relation of this divine absolute infinite to the infinities characterizing his creation, in particular Leibniz's thoughts on the *immensum* and its distinction from the unbounded, on limitation and the unlimited, and Leibniz's distinction in his notes of 1676 on Spinoza's "Letter on the Infinite" among three "degrees" of infinity.[30] For a careful treatment of these questions I refer to the article by Antognazza already cited, as well as two provocative articles by Ohad Nachtomy, in which he tentatively interprets the second of these three degrees as that pertaining to created things, rather than to quantifiable things or to God.[31]

[30] The highest of "degree [*gradus*]" of the infinite distinguished by Leibniz is "that which contains everything", "the absolutely infinite", and this may be identified without controversy as the hypercategorematic infinite characterizing God; while the lowest, "that which is greater than we can expound by any assignable ratio to sensible things", is the syncategorematic infinite in magnitude, of which Leibniz gives as an instance the area between Apollonius's hyperbola and its asymptote. Between these degrees of infinity is "*everything of its kind*, i.e. that to which nothing can be added, for instance, a line unbounded on both sides" (A VI 3, 282/LLC 115).

[31] Nachtomy (2011) suggests that the middle degree is non-quantitative, and applicable to created substances, in contrast to the hypercategorematic, and the syncategorematic. He takes the latter to be applicable only to "numbers and more generally (though with some qualifications) to quantities and magnitudes" (957–958). (For a more extended treatment, see Nachtomy 2016.) I have argued here that the syncategorematic infinite applies equally to created things, such as bodies' trajectories and *petites perceptions*. Also, I agree with Antognazza (2015, 8) that it is more likely that what Leibniz has in mind with his second degree of infinity is "the infinity of the divine attributes" discussed by Spinoza. As I have discussed in (Arthur 2013b), Leibniz interprets divine *immensity* as the basis of space, and divine *eternity* as the basis of time. These are indeed non-quantitative, as Nachtomy suggests, and that is one reason why they cannot be identified with space and time, as they are by Newton and Clarke, according to Leibniz.

My topic, by contrast, may seem on the face of it to be a superficial dispute about the interpretation of a scholastic term. But I believe that a proper grasp of Leibniz's syncategorematic actual infinite in multitude, as well as of his syncategorematic interpretation of infinitesimals, is of the greatest consequence for a correct understanding not only of the foundations of Leibniz's mathematics, but also of his metaphysics of monads and their relation to problem of the composition of the continuum.

References

Antognazza, M. R. (2015). The Hypercategorematic Infinite. *Leibniz Review, 25*, 5–30.

Arthur, R. T. W. (2008). Leery Bedfellows: Newton and Leibniz on the status of infinitesimals. In U. Goldenbaum & D. Jesseph (Eds.), *Infinitesimal differences: Controversies between Leibniz and his contemporaries* (pp. 7–30). Berlin: De Gruyter.

Arthur, R. T. W. (2011). Presupposition, aggregation, and Leibniz's argument for a plurality of substances. *Leibniz Review, 21*, 91–116.

Arthur, R. T. W. (2013a). Leibniz's syncategorematic infinitesimals, smooth infinitesimal analysis, and second order differentials. *Archive for History of Exact Sciences, 67*, 553–593.

Arthur, R. T. W. (2013b). Leibniz's theory of space. *Foundations of Science, 18*(3), 499–528.

Bosinelli, F. C. M. (1991). Über Leibniz' Unendlichkeitstheorie. *Studia Leibnitiana* Bd. 23 H. 2, 151–169.

Breger, H. (1986). Leibniz, Weyl und das Kontinuum. In A. Heinekamp (Ed.), *Beiträge zur Wirkungs- und Rezeptionsgeschichte von Gottfried Wilhelm Leibniz (Studia Leibnitiana Supplementa 26)* (pp. 316–330). Stuttgart: Franz Steiner.

Breger, H. (1990). Das Kontinuum bei Leibniz, 53–67 in Lamarra; reprinted in (Breger 2016a), pp. 115–126.

Breger, Herbert. (1992). Le continu chez Leibniz, 76–84 Le labyrinth du continu, Jean-Michel Salanskis Hourya Sinaceur Paris: Springer; reprinted in (Breger 2016a), pp. 127–135.

Breger, H. (2008). Leibniz's calculation with compendia, 185–198 in (Goldenbaum and Jesseph 2008); reprinted in (Breger 2016a), pp. 147–158.

Breger, H. (2016a). In W. Li (Ed.), *Kontinuum, analysis, informales—Beiträge zur Mathemathik und Philosophie von Leibniz*. Berlin-Heidelberg: Springer.

Breger, H. (2016). On the grain of sand and heaven's infinity. In W. Li (Ed.), *Für unser Glück oder das Glück anderer* (pp. 63–96). Hildesheim-Zürich-New York: Georg Olms Verlag. Hildesheim-Zürich-New York.

Goddu, A. (1984). *The physics of William of Ockham*. Leiden and Cologne: Brill Academic Publishing.

Ishiguro, H. (1990). *Leibniz's philosophy of logic and language* (2nd ed.). Cambridge: Cambridge University Press.

Jesseph, D. (1998). Truth in fiction: Origins and consequences of Leibniz's Doctrine of infinitesimal magnitudes. *Perspectives on Science, 6*, 6–40.

Jesseph, D. (2008). Leibniz and the foundations of the calculus: The question of the reality of infinitesimal magnitudes. In U. Goldenbaum & D. Jesseph (Eds.), *Infinitesimal differences: Controversies between Leibniz and his contemporaries* (pp. 215–233). Berlin: De Gruyter.

Knobloch, E. (2002). Leibniz's rigorous foundation of infinitesimal geometry by means of Riemannian Sums. *Synthese, 133*, 59–73.

Leibniz, G. W. (1969). *Philosophical papers and letters* (L. Loemker, Ed. and Trans.). 2nd. ed. Dordrecht: D. Reidel, Cited as L.

Leibniz, G. W. (1971). In C. I. Gerhardt (Ed.), *Leibnizens Mathematische Schriften*. Berlin: Asher and Schmidt, 1849–1863. Reprint Hildesheim: Georg Olms, 7 vols. Cited as GM.

Leibniz, G. W. (1978). In C. I. Gerhardt (Ed.), *Die Philosophische Schriften von Gottfried Wilhelm Leibniz*. Berlin: Weidmann, 1875–1890. Reprint Hildesheim/New York: Georg Olms, 7 vols. Cited as GP.

Leibniz, G. W. (1981). *New essays on human understanding* (P. Remnant & J. Bennett, Ed. and Trans.). Cambridge: Cambridge University Press, Cited as NE.

Leibniz, G. W. (2007). *The Leibniz-Des bosses correspondence* (selected, edited and translated, with an introductory essay: Look, B., Rutherford, D.). New Haven: Yale University Press, Cited as LDB.

Leibniz, G. W. (2011). *The Labyrinth of the Continuum: Writings of 1672 to 1686* (selected, edited and translated, with an introductory essay: Arthur, R. T. W.). New Haven: Yale University Press, Cited as LLC.

Leibniz, G. W. (2013). *The Leibniz-De Volder correspondence* (translated, edited and with an Introduction: Lodge, P.). New Haven: Yale University Press, Cited as LDV.

Leibniz, G. W. (1923). In Akademie der Wissenschaften der DDR (Ed.), *Sämtliche Schriften und Briefe*. Darmstadt: Akademie-Verlag; cited by series, volume and page, as A VI 2, 123, etc.

Levey, S. L. (2008). Archimedes, infinitesimals, and the law of continuity: On Leibniz's fictionalism. In U. Godenbaum & D. Jesseph (Eds.), *Infinitesimal differences: Controversies between Leibniz and his contemporaries* (pp. 107–133). Berlin: De Gruyter.

Marshall, D. B. (2011). Leibniz: Geometry, physics, and idealism. *Leibniz Review, 21*, 9–32.

Moore, A. W. (1991). *The infinite*. London: Routledge.

Nachtomy, O. (2011). A tale of two thinkers, one meeting, and three degrees of infinity: Leibniz and Spinoza (1675–8). *British Journal for the History of Philosophy, 19*(5), 935–961.

Nachtomy, O. (2016). Infinite and limited: On Leibniz's view of created beings. *Leibniz Review, 26*, 179–196.

Rabouin, D. (2015). Leibniz's rigorous foundations of the method of indivisibles. In V. Julien (Ed.), *Seventeenth-century indivisibles revisited* (pp. 347–364). Switzerland: Springer.

Sereda, K. (2015). Leibniz's relational conception of number. *Leibniz Review, 25*, 31–54.

Uckelman, S. L. (2015). The logic of categorematic and syncategorematic infinity. *Synthese, 192*, 2361–2377.

Chapter 11
The Infinite Given Magnitude and Other Myths About Space and Time

Paul Guyer

Abstract I argue that Kant's claim in the "Transcendental Aesthetic" of the *Critique of Pure Reason* that space and time are immediately given in intuition as infinite magnitudes is undercut by his general theory of mathematical knowledge. On this general theory, pure intuition does not give objects of any determinate magnitude at all, but only *forms* of possible objects. Specifically, what pure intuition itself yields is the recognition that any determinate space or time is part of a larger one, but it requires an inference of reason to go from that to the claim that space and time are infinite. I further argue that this result is consistent with Kant's claim in the second-edition "Transcendental Deduction" that the unity of space and time are the products of synthesis, but also means that the unity of space and time as objects cannot be used a *premise* in the Deduction but can only be regarded as a *conclusion* of the deduction and the following "System of Principles."

11.1 Introduction

In the Transcendental Aesthetic, Kant appears to infer the fourth proposition of the "Metaphysical exposition" of the concept of space, as it is renamed in second edition of the *Critique of Pure Reason*, namely, that "Space is represented as an infinite **given** magnitude," from the preceding proposition, that "Space is not a discursive or, as is said, general concept of relations of things in general, but a pure intuition" (A 25/B 39); similarly with regard to the case of time (A 31–2/B 47–8).[1] His

[1]All quotations from the *Critique of Pure Reason* are from Immanuel Kant, *Critique of Pure Reason*, edited and translated by Paul Guyer and Allen W. Wood (Cambridge: Cambridge University Press, 1998). Quotations from Kant's inaugural dissertation, *On the Forms and Principles of the Sensible and Intelligible Worlds* ("*ID*") are from Kant, *Theoretical Philosophy 1755–1770*, translated and edited by David Walford in collaboration with Ralf Meerbote (Cambridge: Cambridge University Press, 1992).

P. Guyer (✉)
Philosophy Department, Brown University, Providence, RI, USA
e-mail: paul_guyer@brown.edu

O. Nachtomy, R. Winegar (eds.), *Infinity in Early Modern Philosophy*, The New Synthese Historical Library 76,
https://doi.org/10.1007/978-3-319-94556-9_11

argument, although not entirely explicit, would seem to be quite straightforward. But I will argue here that it runs afoul of Kant's own view, clearly expressed elsewhere in the *Critique*, that pure intuition does not provide determinate representations of actual objects, but only representations of *forms* of possible objects that are determinate in some regards but indeterminate in other regards. Indeed, I will argue, Kant's claim that pure intuition presents a representation of a *single* token of space or time, thus that space and time are not only the pure forms of intuition but themselves intuitions in the sense of representations of singular objects, is problematic. This result will be of obvious importance for the interpretation of Kant's philosophy of mathematics.

But it may also be important for the interpretation of the larger argument of the *Critique*. Henry Allison has recently argued that the second-edition version of the transcendental deduction of the categories culminates in the inference from the premise that space and time are each represented as a single object in pure intuition through the lemma that the cognition of a manifold of representations as a manifold presenting a single object requires the use of the categories to the desired conclusion that all empirical representations of objects in space and time are subject to the categories. But if the pure intuitions of space and time are merely representations of possible forms of objects and do not by themselves suffice for the representation of singular space and time, the unity of such a world of empirical objects determinately located and related in space and time under the categories cannot be inferred from the pure form of spatio-temporal representation but must be independently established. Reflection on why by Kant's own lights the pure intuitions of space and time cannot suffice for the determinate representation of them as infinite and singular given magnitudes requires an alternative to Allison's interpretation of the transcendental deduction, or even, if Allison's interpretation does faithfully represent a line of Kant's own thought, an alternative to Kant's own presentation of the argument.

11.2 Are Space and Time Given as Infinite and Singular Magnitudes in Pure Intuition?

In the Transcendental Dialectic of the *Critique of Pure Reason*, in his 'explanation of the dialectical illusion in all transcendental proofs of a necessary being" (A 614/B 642), Kant himself warns us that the notion of space as an "*a priori* object given in itself" is an hypostatization and "transcendental subreption" from what is a mere form or "principle of sensibility." Indeed, he takes the difference between a mere form of representation and an actually given object to be so obvious that he can use the case of space to buttress his argument that the "ideal of a highest being" is nothing other than a "regulative principle of reason," not an actual object of theoretical cognition:

> The ideal of a highest being is. . . nothing other than a **regulative principle** of reason, to regard all combination in the world **as if it** arose from an all-sufficient necessary cause, so as to ground on that cause the rule of a unity that is systematic and necessary according to universal laws; but it is not an assertion of an existence that is necessary in itself. But at the same time it is unavoidable, by means of a transcendental subreption, to represent this formal principle to oneself as constitutive, and to think of this unity hypostatically. For, just as with space, since it originally makes possible all forms which are merely limitations of it, even though it is only a principle of sensibility, it is necessarily held to be a Something subsisting in itself and an *a priori* object given in itself, so it also comes about entirely naturally that since the systematic unity of nature cannot be set up as a principle of the empirical use of reason except on the basis of the idea of a most real being as the supreme cause, this idea is represented as an actual object, and this object again, because it is the supreme condition, is represented as necessary, so that a **regulative** principle is transformed into a **constitutive** one ... (A 619–20/B 647–8)

According to Kant, it is a sound regulative principle to seek the kind of unity among laws of nature that they would have if they were grounded in a supreme being, but a natural yet theoretically unwarranted hypostasis to assert that such a being actually exists; likewise, it is necessary to represent any space as part of a larger one, but this is only the *a priori* form of our (outer) sensibility, not based on any "*a priori* object given in itself" as "a Something subsisting in itself": it is a mistake, a "subreption" in Kant's term, to confuse the indefinitely extendable form of (outer) intuition with any actually given infinite magnitude.

Yet in the Transcendental Aesthetic of the *Critique*, in the four arguments that he labeled in the second edition of the book the "Metaphysical Exposition" of the concept of space" (B 37), Kant had plainly asserted that "Space is represented as an infinite **given** magnitude." His basis for this assertion lies in the preceding argument for the thesis that our representation of space, already shown to be *a priori* because it is presupposed by our empirical representations of any particular objects in space, is not a "discursive" or "general concept" but a "pure intuition." Here are the preceding argument and then Kant's inference from it:

> 3) Space is not a discursive or, as is said, general concept of relations of things in general, but a pure intuition. For, first, one can only represent a single space, and if one speaks of many spaces, one understands by that only parts of one and the same unique space. And these parts cannot as it were precede the single all-encompassing space as its components (from which its composition would be possible), but rather are only thought **in it**. It is essentially single; the manifold in it, thus also the general concept of spaces in general, rests merely on limitations.

After an apparent digression in which Kant appeals to his doctrine that a synthetic *a priori* geometrical proposition such as "in a triangle two sides together are always greater than the third" can never be proved through the analysis of concepts but can only be demonstrated by construction in pure intuition, which would seem to belong better in the transcendental exposition of space, which argues that pure intuition is the condition of possibility of synthetic *a priori* cognition in geometry, Kant returns to the main argument and draws the conclusion that

> 4) Space is represented as an infinite **given** magnitude. Now one must, to be sure, think of every concept as a representation that is contained in an infinite set of different possible

> representations (as their common mark), which thus contains these **under itself**; but no concept, as such, can be thought as if it contained an infinite set of representations **within itself**. Nevertheless space is so thought (for all the parts of space, even to infinity, are simultaneous). Therefore the original representation of space is an ***a priori*** **intuition**, not a **concept** (B 39–40).

I will be focusing on the case of space and spatial forms in this section, but since Allison's interpretation of the transcendental deduction, to which I will turn in the second part of the paper, foregrounds the role of the representation of time, before discussing what has just been quoted I will also quote Kant's parallel argument concerning the representation of time. Here Kant writes:

> 4) Time is no discursive, or as one calls it, general concept, but a pure form of sensible intuition. That representation, however, which can only be given through a single object, is an intuition. Further, the proposition that different times cannot be simultaneous cannot be derived from a general concept. The proposition is synthetic, and cannot arise from concepts alone. It is therefore immediately contained in the intuition and representation of time.

From this Kant infers that

> 5) The infinitude of time signifies nothing more than that every determinate magnitude of time is only possible through limitations of a single time grounding it. The original representation **time** must therefore be given as unlimited. But where the parts themselves and every magnitude of an object can be determinately represented only through limitation, there the entire representation cannot be given through concepts (for they contain only partial representations), but immediate intuition must ground them. (B 47–8).

(The two propositions are numbered "3" and "4" in the second-edition exposition of the concept of space but "4" and "5" in the exposition of time. This is because in the first case Kant has removed the preceding argument (3) from the synthetic *a priori* status of geometrical propositions to their basis in space as the pure form of outer intuition to a separate "transcendental exposition" of the concept of space (although, as we saw, he also left a reference to this argument in the newly renumbered (3)), but in the case of time he has left in place the argument (3) that the pure form of time "also grounds the possibility of apodictic principles of relations of time" and merely repeated that point in the now separate "transcendental exposition of the concept of time" (B 47–8).)

In neither case is the ground for Kant's inference from the first proposition to the second actually explicit. The argument about space first claims that one can represent only a single space because multiple spaces can only be represented as parts of a single space encompassing them all, as so to speak carved out of the larger space. If regions of space were given as parts pre-existing the whole of space, then of course the magnitude of space as a whole would depend upon both the magnitude and number of the parts antecedently given and a completed synthesis of them, and a claim to the infinitude of the whole of space would depend upon an infinite number of parts being given and synthesized into a whole, which is impossible for us. In that case perhaps even the unity or singularity of overall space would depend on whether the pre-given parts were all combinable into one. But if the parts of space are only carved out of a pre-given whole, then the neither the singularity nor the infinitude of the latter would depend on the number, magnitude, and synthesis of the individual spaces. Still, Kant does not quite explain how the positive assertion that "Space is represented as an infinite **given** magnitude" follows. Instead, having asserted this conclusion, Kant just draws a contrast with the case of concepts: a concept cannot contain an infinite number of parts, namely "marks" or predicates, although it can be predicated of an infinite number of instances of those marks; but the representation of space actually has an infinite number of parts, namely smaller spaces, and thus cannot be a concept at all, but an intuition, a singular representation, that is, representation of something singular, and because everything asserted thus far is *a priori* rather than empirical, a pure intuition. In the case of time, Kant states this contrast first, in proposition (4), maintaining that because "Different times are only parts of one and the same time" the representation of time as a whole must be singular and thus an intuition, not a general concept, which, although he does not repeat this, leaves the number of its instances open. He then adds, as proposition (5), that the "infinitude of time" is equivalent to ("signifies nothing more than") "that every determinate magnitude of time is only possible through limitations of a single time grounding it," and that "The original representation **time** must therefore be given as unlimited." How the second claim follows from the first is again not made explicit. Why does the fact that any particular space or time is represented as a region of a larger one imply that space and time are *infinite* given wholes?

What is missing from both expositions must be the supposition that this analysis holds for any region of space or time as a part of a larger space or time *no matter how large the former*. If any region of space or time *no matter how large* must be represented of as part of a yet larger space or time, then it would follow that the latter must be "unlimited" or infinite in extent or duration. The whole argument would then be something like this: Particular regions of space or durations of time are not given as parts preceding the whole of space or time, in which case the magnitude of space or time as a whole and perhaps even their number would depend upon the magnitude and combinability of the parts. Instead, particular spaces or times are only regions carved out of the whole of space or time by the introduction of boundaries or limits within the whole. But further, if this is true of any particular space or time no matter how large, then the whole of space or time must always be represented as larger, thus as unlimited or infinite. The representations of both particular spaces and times as well as the representations of space and time as a whole are intuitions, that is, singular representations, the former always finite in extent but the latter infinite because no matter how big the former are the latter is always bigger. This contrasts with the case of concepts, Kant infers, for concepts themselves are always finite, that is, composed of a finite number of marks or predicates, even if they can have an infinite number of instances – although he should say only that concepts leave the number of their instances undetermined and thus admit of an *indefinite* number of instances. Note that it follows from this argument that the *concepts* of *particular* spaces or particular times do have an infinite – or indefinite – number of instances because space and time are themselves represented as infinite and thus as delimitable into an infinite number of parts, even though the representations of both space and time as wholes and those of their delimited parts are themselves intuitions. That is, I can have the concepts of a line a foot long or of a duration an hour long, and such concepts of course have instances. They can indeed be supposed to have an infinite number of instances – but only because the underlying intuitions of space and time as such are supposed to be infinite.

My claim is thus that Kant's first point, that space (or time) is represented as singular, depends on the premise that any particular space (or time) is represented as part of a larger one, and his second point, that space (or time) is represented as infinite, depends on the further assumption that the first premise is true no matter how large any particular space (or time) contained in that larger one is. That this is the structure of Kant's reasoning is more explicit in his draft comments on an essay by the mathematician Abraham Kästner that had been published in J.A. Eberhard's *Philosophisches Magazin*, the locus for Eberhard's attack upon Kant's conception of synthetic *a priori* cognition, especially in mathematics. In these notes, Kant largely endorses Kästner's own response to Eberhard, for he sees Kästner as agreeing with his own position that the justification of mathematical concepts requires not just logical consistency, as the Leibnizian Eberhard believed, but also "rules of connection" derived from "pure intuition" (20,412). But the passage that concerns us here is this:

> When in behalf of metaphysics it is said that the space originally given in our power of representation is infinite this means nothing more than: all spaces that may be given are possible only as parts of a singular one [*eines eintzigen*][, that is,] can be given only as parts belonging to a single [*einigen*] space …; a space, however, of which <u>any space</u> that is determinate in size is only a part is greater than <u>any</u> *quantum spatii spatium dabile*, that is, is greater <u>than any</u> space that I can describe and that means that it is infinite. This infinitude, which one can call merely metaphysical (i.e., subjective in the form of our sensibility but not objective outside of that and in the content of things in themselves), is entirely real with regard to all objects of our outer sensible intuition … For that one can continue a line infinitely [*ins Unendliche*] or add planes to one another <u>so far as one will</u>, this potential infinitude on which alone the mathematician has to ground all of his determinations of space presupposes this actual (but only metaphysical) infinitude and is possible only under this presupposition. For what does it mean that <u>no matter how far one has drawn a straight line one can always draw it yet further?</u> It means that the space in which I describe a line is greater than any space that I might describe within it. (20:418, emphasis added)

This passage clearly begins with the premise that any space, or spatial determination, any line or plane, must be represented as part of a larger one, but then pretty clearly adds, by means of the various phrases I have underlined, that this remains true no matter how large such a spatial determination is. But if the containing space is always larger than the contained one, no matter how large the latter is, then, Kant infers, the containing space must be infinite in size. This is the inference that Kant is making to go from point (3) to (4) in the second-edition metaphysical exposition of space (or from (4) to (5) in the case of time).[2]

[2] I believe that the Kästner comments answer the objection made to me in comments on an earlier draft of this paper by both Rosalind Chaplin and Lisa Shabel, who each take the claim in (4) of the B-edition metaphysical exposition of space that space is an infinite given magnitude to be an independent and self-evident premise for the conclusion that the representation of space is an intuition

I will come back to the question of what Kant means by saying that the mathematician's determination of space presupposes a space that is "merely" or "only" metaphysical at the end of this section. But first I want to discuss what I consider to be serious problems with Kant's line of thought as I have reconstructed it. The most obvious objection is that Kant does not actually show that our representation of space as unlimited or infinite is immediately *given*; rather, the infinitude of space seems to be *inferred* from the fact that any particular space can be represented only as delimited from a larger surrounding space. We might regard what is immediately given in intuition as merely the intuition that *some* represented space is surrounded by one that is larger than *that*, but that space must be represented as infinite would seem to follow only from the *thought* that such an intuition of one space as enclosed in a larger one *can be reiterated*, or alternatively from the *reflection* that the same sort of intuition of smaller space within larger space could be had no matter how large the smaller space, leading to the *conclusion* that space as a whole must be able to be represented as infinite because it can be represented as larger than any smaller space no matter how large. It seems implausible to describe such an inferred or reflective conclusion as immediately intuited, thus implausible to describe the infinite magnitude of space as immediately *given*.

Kant himself seems to indicate that the representation of any infinite magnitude is only inferential, indeed a product of *reason*, in his commentary on the Antinomy of Pure Reason. Here he writes that pure reason

> cannot say **what the object is**, but only **how the empirical regress is to be instituted** so as to attain to the complete concept of the object. For if the former were the case, then it would be a constitutive principle, the likes of which is never possible on the basis of pure reason. Thus with it one can by no means have the intention to say that the series of conditions for a given conditioned is in itself finite or infinite; for in that way a mere idea of the absolute totality, which is produced only in the idea itself, would think an object that cannot be given in any experience ... Thus the idea of reason will only prescribe a rule to the regressive synthesis in the series, a rule in accordance with which it proceeds from the conditioned, by means of all the conditions subordinated to one another, to the unconditioned, even though the latter will never be reached. For the absolutely unconditioned is not encountered in experience at all. (A 510/B 538)

Now it will immediately be objected that this statement concerns the *empirical* synthesis of a series of objects in space and time, not the pure intuition of space and time themselves, and further that even the *possibility* of an unending regress or progress of synthesis of representations of objects in space and time depends on the antecedent *intuition* of space and time themselves as infinite, thus permitting such

rather than a concept; the Kästner comment shows that Kant derives this premise from the containment thesis of point (3) plus the assumption that the containment thesis applies to a space of any size. In support of this interpretation it might also be noted that in the inaugural dissertation, Kant's first version of the Transcendental Aesthetic, he lumps the containment-thesis and the boundlessness-thesis together in a single argument (*ID*, §15.B, 2:402 in the case of space, §14.2, 2:399 in the case of time). I would like to thank both Shabel and Chaplin for their comments.

regress or progress.[3] But Kant immediately illustrates his argument with a purely geometrical example:

> One can rightly say of a straight line that it could be extended to infinity, and here the distinction between the infinite and a progress of indeterminate length (*progressus in indefinitum*) would be an empty subtlety. For although when it is said "Draw a line" it obviously sounds more correct to add *in indefinitum* than if it were said *in infinitum*, because the first means no more than "Extend it as far as you **want**," but the second means "You **ought** never to stop extending it" … yet if we are talking only about what **can** be done, then the first expression is entirely correct, for you could always make it greater, to infinity. And this is also the situation in all cases where one is speaking only of a forward progress … For here reason never needs an absolute totality in the series, because it is not presupposed as given (*datum*), but it is only added on as something conditioned, which is capable of being given (*dabile*), and this without end. (A 511–12/B 539–40)

I am not so sure that the difference between *in indefinitum* and *in infinitum* is a subtlety, and am inclined to think that just as Kant should have said that a concept leaves the number of its instances indefinite so he should have said that space and time are represented as indefinitely large rather than infinite. But my main point is that the representation of space and time as either indefinite or infinite *depends upon an inference*, an act of reason, to the conclusion that the relevant kind of series, whether it is the extension of a line or a duration or the encasement of a region or duration within a larger one, can always be continued, while pure intuition provides only the *dabile*, not the *datum*, that is, it makes the series *extendable* but does not itself accomplish the extension. The latter is something inferred by reason, and indeed is like an *idea* of reason, which is precisely the idea of something that is *never actually given.* On this account, the pure form of intuition – the intuition that any space or time is contained in a larger one – provides the *ground* for an *inference* that the representation of space and time can therefore be infinitely extended, but no representation of space and time as actually infinite is ever *given*, whether by intuition or reason.

Even those who accept Kant's claim that the infinite magnitude of space and time is given in pure intuition actually seem to presuppose an inference to infinite magnitude. Thus, Colin McLear, who wants to defend the claim that space (he does not consider time) is given as both infinite and singular in pure intuition, prior to any conceptually-grounded synthesis (the position he calls "Sensibilism" rather than "Intellectualism"), writes that "any perceptual experience of a space, or of something shaped and located in space, requires a representation of that space as merely a limitation of a larger whole. So for any experience of a bounded region, the experience is made possible in virtue of a representation of a larger region that encompasses the bounded one, and so on."[4] His use of the phrase "and so on," it seems to me, implies that the infinitude of space is inferred from the possibility of reiterating the "experience" of any region of space as encompassed by a larger one. Similarly,

[3] See Michael Friedman, "Kant on Geometry and Spatial Intuition," *Synthese* 186 (2012): 231–55, at pp. 240–1, and his quotation there from Kant's response to Eberhard at *Ak*. 20: 420–1.

[4] Colin McLear, "Two Kinds of Unity in the *Critique of Pure Reason*," *Journal of the History of Philosophy* 53 (2015): 79–110, at p. 95.

Michael Friedman, who argues against a "diagrammatic" interpretation of Kant's philosophy of geometry, according to which the universal claims of geometry are inferred from the construction of particular diagrams by abstracting from their metrical rather than topological properties (as well as from their imprecision), that the possibility of such abstraction rests on an underlying intuition of the pure form of space within which the constructions represented by diagrams (when they are) takes place, presupposes an essential operation of *iteration* that is intellectual rather than purely intuitional in character: "Reasoning by substitution is…essentially *iterative*, and it is precisely such iterative thinking, for Kant, that underlies both pure geometry (in the guise of Euclidean proof) and the more general calculative manipulation of magnitudes in algebra and arithmetic."[5] Whatever faculty he thinks performs this iteration, he seems blocked from the view that the infinite magnitude of space is simply given in pure intuition: from pure intuition it seems to be a *dabile* but not a *datum*.

Like McLear, Christian Onof and Dennis Schulting also want to defend the claim that there is a kind of unity of space given prior to any synthesis by the understanding or intellect. To avoid conflict with Kant's frequent claims that unity always depends on synthesis, they introduce the term "unicity" to characterize this preconceptual unity of space, which they hold to be comprised by its "singularity," that "one can only represent a single, unique space," its "mereological inversion," that "the whole of space precedes its parts," and its "infinity," that the "magnitude of space is given."[6] However, they then quickly quote Kant as stating in a reflection that "the capacity to receive several impressions of outer objects, or the susceptibility to them, has no limits in itself."[7] This pretty clearly says that the infinitude of space is *dabile* but not *datum*. They go on to argue that "if the understanding were to produce space by affecting sensibility, it is not clear how an actual infinity is a possibility here: the understanding can certainly have no grasp of such an infinity, let alone be responsible for the imagination's ability to generate it"[8] But while this is correct if taken to mean that no inference whether of the understanding or reason can ever yield the existence of an actual infinitude, or exhibit it, it does not deny that the idea of the infinitude of space is a thought that must somehow be generated from the pure intuition of any particular space as contained in a larger one – even if this idea is ascribed to the imagination, it is not simply intuited. My argument is that no single intuition of the form of space can present us with its actual infinitude, rather that this idea can be generated only by an intellectual act of reiterating the image of

[5] Friedman, "Kant on Geometry and Spatial Form," pp. 237–8.

[6] Christian Onof and Dennis Schulting, "Space as Form of Intuition and as Formal Intuition: On the Note to B 160 in Kant's *Critique of Pure Reason*," *Philosophical Review* 124 (2015): 1–58, at p. 15.

[7] Kant, Reflection 4673, 17:641; from Immanuel Kant, *Notes and Fragments*, edited by Paul Guyer, translated by Curtis Bowman, Paul Guyer, and Frederick Ruascher (Cambridge: Cambridge University Press, 2005), pp. 156–7.

[8] Onof and Schulting, "Space as Form of Intuition and as Formal Intuition," p. 19. They make this claim in criticism of Béatrice Longuenesse.

a space enclosed in a larger one without limit, although of course that idea is not the same as an actual synthesis of infinite space, which can never be completed. In Kant's own terminology, as we have seen, it is more like pure reason's idea of the unconditioned, which is never presented in a completed synthesis.[9]

A second, perhaps less obvious objection, but one that will turn out to be even more important in the subsequent discussion of the role of the pure intuition of space and time in the transcendental deduction, is that different instances of pure intuition do not obviously present the *same single object*, but only the *same form* of objects. If I prove one theorem about triangles in pure intuition, say the one just discussed, by constructing a triangle in pure intuition, and then prove a different theorem about triangles by means of another construction in pure intuition, I am not representing the *same object* on two different occasions, but the *same form of objects*, namely, the triangular form, or the triangle as form. Representing the form of a triangle twice is not like representing my desk before me now, at 10 AM, then representing the same desk at 11 AM when I come back from the kitchen with a fresh cup of coffee, and then again when at 1 PM when I come back from lunch. In that case I am obviously having repeated representations of one and the same physical object. But when I represent a triangle in pure intuition for the purpose of proving one theorem, then represent a triangle again for the purpose of proving another theorem, or perhaps even for proving the same theorem again, there is not one,

[9] In this connection it could also be pointed out that in at least some regards what is represented in pure intuition does not have any determinate magnitude at all. We can see this by considering one of Kant's typical examples of the representation of a geometrical figure in pure intuition, namely the construction of a triangle in pure intuition in order to prove that the sum of the interior angles of any triangle equals 180° (A 716/B 744). Kant asks us to imagine a triangle and to imagine extending one of its sides and then dividing the external angle thus created with another line parallel to the opposite side of the triangle; we will then see that the two angles thus created plus the adjoining internal angle together occupy one side of the straight line constructed by previously extending the side of the triangle, thus that they equal 180°; but we will also see that each of the two newly divided angles is equal in size to one of the other original interior angles, because the line dividing the two exterior angles is parallel to the opposite side of the original triangle, from which it follows that the original interior angles also add up to 180°. Now, one could argue that in this case too the result is not immediately given, but is rather inferred from or on the basis of what is intuited, but that is not the point I want to emphasize. Rather, the point is that while in this example *one* (and in fact only one) magnitude is determined, namely that the sum of the interior angles of a triangle equals 180° (although what the magnitude of each of the three individual interior angles might be is left undetermined), *other* magnitudes are left undetermined, in particular the *length* of the sides of the triangle and thus the size of the triangle as a whole is left undetermined – as indeed must be the case if the proof is to be valid for *all* triangles. And this suggests that pure intuition never gives objects of any determinate size, although it can present one figure as contained in another, thus the latter as larger than the former. This in turn suggests that the pure intuition of regions of space as contained in a larger space does not present the former as having any determinate size, thus neither does it present the latter as having any determinate size; it presents it merely as indeterminately larger than the former. This puts the burden of insight into the unlimitedness of space as a whole back onto the idea of iteration, that is the recognition that the intuition of a smaller space as contained within a larger one can be indefinitely reiterated, and thus that the larger space must be unlimited. But this just reinforces the objection that the infinite magnitude of space is inferred, not simply given in pure intuition.

continuously existing triangle at a determinate location that I am representing, the way I am representing one desk continuously existing at one spot in my room, at one intersection of latitude and longitude on the surface of the earth, which is in turn at a determinate distance from the sun this morning, and so on; I am repeatedly representing the same form, but not the same determinate object. And this is because in representing the form of a triangle in pure intuition, I am not representing a determinate object at all. I am just representing a form that may be determinate in certain regards but not in others, but in any case not a fully determinate actual object like my desk. Contrary to what Onof and Schulting assert to be the pre-synthetic "unicity of space" and time, what McLear asserts to be its pre-conceptual unity, and what Friedman, following Kant himself in the Eberhard controversy, asserts to be the singularity of space presupposed by geometry,[10] pure intuition does not present a single space and a single time, but a single form of space and time. When I imagine constructing geometrical figures, I am not locating them determinately in some larger but still pure intuition of space. If I construct a triangle in pure intuition, it has no determinate location in any larger space, even though I can always imagine extending any one of its sides for the sake of a proof. And while if I construct, whether in imagination or on paper, as Kant says, one figure within another, then obviously the former is contained in the imagined space of the latter, if I construct two different figures on two different occasions, even if I draw them on the same sheet, then, as pure forms of space, they have no determinate relation to one another. If I construct the pure figure of an isoceles triangle and then the pure figure of a right triangle, those pure figures are not six inches apart from each other, even if the drawings I use to help me imagine them are. In geometry, at least, there is no such thing as the actual whole of pure space, just the possibility of imagining spatial constructions, including the encasement of any figure in a larger one. Onof's and Schulting's claim that "our science of geometry always operates with the assumption of a unique background space…a priori spatial representation can only ground geometrical constructions insofar as such constructions can be carried out without requiring different spaces"[11] is false: my proof of one theorem about triangles in (for example Euclidean) space by means of one construction of a triangle and my proof of another by construction of another figure do not presuppose any determinate relation between two triangles in one space; they presuppose only that they are constructions in the same *kind* of space.[12]

To be fair, Onof's and Schulting's assumption is not without textual support. In the inaugural dissertation, Kant links the unity of space to everything represented in

---

[10] Onof and Schulting, "Space as Form of Intuition and Formal Intuition," p. 15; McLear, "Two Kinds of Unity," pp. 86–93; Friedman, "Kant on Geometry and Spatial Form," pp. 240–1.

[11] Onof and Schulting, "Space as Form of Intuition and as Formal Intuition," p. 21.

[12] Onof and Schulting observe that "there is no location specificity to constructions in space: when I construct a triangle, there is no reference to where it is constructed"; "Space as Form of Intuition and as Formal Intuition," p. 40. That is correct, and it means that when I construct another triangle in pure intuition there is no determinate spatial relation between the two triangles; thus Onof and Schulting's prior assertion that geometry requires a single (token of) space is false.

it being represented in a determinate relation to everything else: "For what you speak of as *several places* are only parts of the same boundless space related to one another by a fixed position" (*ID*, §15.B, 4:302). The problem is just although in each case of representing a space in pure intuition I immediately represent that space as part of a larger space, in multiple instances of the pure intuition of space I do not immediately represent the spaces represented (either the contained or the containing space) as in any determinate relation to each other, and thus do not immediately represent a single, boundless space. Each time I represent space or time in pure intuition, whether in constructing a figure or demarcating a region from a larger one, I represent a single space or time, and every time I represent space or time in pure intuition I represent it as having the same form, but there is no determinate spatial or temporal relation between the figures or sequences I construct on different occasions (though there are determinate spatio-temporal relations between *my acts of representing* them), thus they are not part of some single continuing space or time. This point is going to be important for the discussion of the role of the pure intuitions of space and time in the transcendental deduction because it means that numerically distinct pure intuitions of the space and time that encompass particular regions of space and time are not representations of the same, actually existing and enduring space and time, *a fortiori* not representations of a single, enduring, all-encompassing space or time unifying all empirical intuitions and their objects, but only representations of the same form, the forms of unlimited space and time. Whether these forms represent single realities (space and time) in which all lesser spatio-temporal realities (spaces and times) can and must be united is a further question, which must be settled if at all on other grounds. And that means that the Transcendental Deduction will not be able to prove that all empirical intuitions can be synthesized into a single manifold of objects under the categories on the basis of a premise of the unity of all space and time as singular already established by the Transcendental Aesthetic, because that argument concerns only the form of intuitions of space and time, not single objects, actual space and actual time, represented by multiple intuitions with spatial or temporal form.

I will develop this point more fully in the second part of the paper. But first I want to forestall criticism that my argument with Kant's conclusions in the metaphysical expositions of space and time that pure intuition immediately gives or presents space and time themselves as single and unlimited, thus my denial that the pure forms of intuition are themselves intuitions of singular objects, is based on non-Kantian assumptions. My argument has turned above all on the claim that pure intuition does not present representations of determinate objects at all, but only representations of forms of objects, or, we might say, that it does not present actual objects, but only the forms of possible objects. I take it that this is precisely what Kant himself says in the second edition of the transcendental deduction. In §§22 and 23 of the revised text, Kant is carefully describing the double-edged sword that is his doctrine that cognition of an object requires both concepts and intuitions, what Henry Allison has dubbed Kant's "discursivity" thesis.[13] That all knowledge of

[13] See Allison, *Kant's Transcendental Deduction*, pp. 46–7, and *Kant's Transcendental idealism:*

actual objects requires both concepts and intuitions means that concepts alone never suffice for knowledge of actual objects, as Kant will emphasize throughout his critique of traditional speculative metaphysics in the Transcendental Dialectic. But it also means that intuitions alone never suffice for knowledge of actual objects, *neither empirical intuitions nor pure intuitions*. The former have to be synthesized in accordance with concepts, themselves formed in accordance with the categories, in order to yield knowledge of actual, determinate objects, and the latter, pure intuitions, have to be supplemented with both empirical intuition and conceptually-guided synthesis in order to yield knowledge of actual, determinate objects.

Kant makes all of these points in §22. The title of this section asserts that "The category has no other use for the cognition of things than its application to objects of experience," but the section itself makes the further point that pure intuition alone also yields no cognition of things without application to objects of experience. The section thus needs to be quoted at length:

> To **think** of an object and to **cognize** an object are thus not the same. For two components belong to cognition: first, the concept, through which an object is thought at all (the category), and second, the intuition, through which it is given; for if an intuition corresponding to the concept could not be given at all, then it would be thought as far as its form is concerned, but without any object, and by its means no cognition of anything at all would be possible, since, as far as I know, nothing would be given nor could be given to which my thought could be applied. Now all intuition that is possible for us is sensible (Aesthetic), thus for us thinking of an object in general through a pure concept of the understanding can become cognition only insofar as this concept is related to objects of the senses.

Thus far, Kant is expounding the basis for his entire critique of the tradition of speculative metaphysics based on concepts of the understanding, in Kant's view illicitly transformed into ideas of pure reason under the aegis of the idea of the unconditioned (see A305-9/B 362-6), alone. But then he switches to the relation between pure and empirical intuition in order to argue that, like the categories, pure intuitions alone also do not give knowledge of any actual objects. Continuing without interruption from what was just quoted, he adds:

> Sensible intuition is either pure intuition (space and time) or empirical intuition of that which, through sensation, is immediately represented as real in space and time. Through determination of the former we can acquire *a priori* cognition of objects (in mathematics), but only as far as their form is concerned, as appearances; whether there can be things that must be intuited in this form is still left unsettled. Consequently all mathematical concepts are not by themselves cognitions, except insofar as one presupposes that there are things that can be presented to us only in accordance with the form of that pure sensible intuition. **Things in space** and **time**, however, are only given insofar as they are perceptions (representations accompanied with sensation), hence through empirical representation.

Then Kant returns to the subject of the categories, arguing that since pure intuition cannot present us with actual objects by itself but only in conjunction with empirical intuition, the intuition with which the categories must be supplemented in order to yield cognition of actual objects must also be empirical and not just pure intuition:

An Interpretation and Defense, revised edition (New Haven: Yale University Press, 2004), pp. 12–16, 27–8.

> The pure concepts of the understanding, consequently, even if they are applied to *a priori* intuitions (as in mathematics), provide cognition only insofar as these *a priori* intuitions, and by means of them also the concepts of the understanding, can be applied to empirical intuitions. Consequently the categories do not afford us cognition of things by means of intuition except through their possible application to **empirical intuition**, i.e., they serve only for the possibility of **empirical cognition**. This, however, is called **experience**. The categories consequently have no other use for the cognition of things except insofar as these are taken as objects of possible experience. (B 146–8)

What is crucial for my present purpose is the middle part of this section (one long paragraph), for what it claims is precisely that pure intuition does not properly present us with objects at all, but only with the forms of possible experience. The "*a priori* cognitions of objects (in mathematics)" is not properly speaking knowledge of determinate objects at all, or we might even say that the term "mathematical objects" would be a misnomer for Kant; mathematics rather presents objects "only as far as their form is concerned," that is to say, it presents us not with actual objects but only with the forms of possible objects. For example, geometry may present us with right or isoceles triangles, thus with triangles with angles of a determinate size, but it does not present us with sides of determinate length, so it does not present us with determinate triangles. It presents us with the form of possible triangular objects, while only perception – "representations accompanied with sensation" – presents us with actual triangular objects. Likewise, pure intuition does not present us with a single, infinite space and time with actual objects occupying determinate regions determinately related to each other. Rather, *each intuition* of a region of space or time presents that region as part of a larger space or time also containing any other regions represented *in that intuition*; but whether *all* our actual, empirical intuitions of regions of space and time can be unified into the representation of a single, unitary, and unlimited space or time is a further question.

Before I finally turn to that further question, let me quote one more passage to confirm my claim that on Kant's own account pure intuition presents only the form of possible objects and not determinate actual objects. This comes from Kant's contrast between the methods of mathematics and philosophy in the Doctrine of Method's chapter on "The Discipline of Pure Reason." Here Kant writes:

> **Philosophical** cognition is **rational cognition** from **concepts**, mathematical cognition that from the **construction** of concepts. But to **construct** a concept means to exhibit *a priori* the intuition corresponding to it. For the construction of a concept, therefore, a **non-empirical** intuition is required which, consequently, as intuition, is an **individual** object, but that must nevertheless, as the construction of a concept (of a general representation), express in the representation universal validity for all possible intuitions that belong under the concept. Thus I construct a triangle by exhibiting an object corresponding to this concept, either through mere imagination, in pure intuition, or on paper, in empirical intuition, but in both cases completely *a priori*, without having to borrow the pattern for it from any experience. The individual drawn figure is empirical, and nevertheless serves to express the concept without damage to its universality, for in the case of this empirical intuition we have taken account only of the action of constructing the concept, to which many determinations, e.g., those of the magnitude of the sides and the angles, are entirely indifferent,

> and thus we have abstracted from these differences, which do not alter the concept of the triangle. (A 713-14/B 741-2).[14]

A pure intuition can itself be regarded as an "individual object" insofar as it is considered to be a particular mental occurrence, itself presumably happening at some particular point in time and some loosely defined region of space (not, in Kant's view, at a particular point in one's brain, but in the region of one's body),[15] but what it is an intuition *of* is not a particular, actual object, but a form of possible objects to which "many determinations, e.g., those of the magnitude of the sides and the angles, are entirely indifferent." For example, in the proof that the interior angles of any triangle equal 180°, the length of the sides of any triangle are left completely indeterminate, and thus no particular triangular object is determinately represented; and in the proof of the Pythagorean theorem, the sizes of the angles are determinate while the *relative* lengths of the sides (the ration of the hypotenuse to the other two sides) is determined but their length by any other measurement is left undetermined; thus the pure intuition of a right triangle with two equal sides for the purposes of that proof is also not the representation of any actual, fully determinate object.

My suggestion is that the same applies to the pure intuition of space and time themselves: any particular pure intuition of a space or a time will be an intuition of it as a region of a larger space or time, but in pure intuition alone neither the regions nor the larger wholes are determinate objects, multiple instances of pure intuition are not different representations of any single, actual object, and therefore the synthesis of all representations of particular regions of actual space and time into a representation of real space and time as unitary, infinite wholes cannot depend on the pure form of spatial and temporal intuition alone. We should now consider what that implies for the interpretation of the transcendental deduction. Before we do, however, I want to consider a last possible objection to the argument I have been making. This is that Kant sometimes does speak of the space and time of pure intuition as if they were proper objects, namely in the same comments on Kästner that I have used as support for my interpretation of Kant's argument for his claim that space and time are infinite given magnitudes.

In this text, Kant argues that the geometrical construction of spaces presupposes "metaphysical" space. Both the former and the latter sound like objects. But in fact what Kant means by metaphysical space is nothing other than the pure *intuition* of space, or the pure form of such intuition. He writes:

> [That] a line can be continued infinitely means nothing more than that space, in which I describe the line, is greater than any line that I may describe in it; and so the geometer grounds the possibility of his task, to expand a space (of which there are many) infinitely,

[14] This passage could seem to support the "diagrammatic" interpretation of Kant's philosophy of geometry by Kenneth Manders that Friedman attacks in "Kant on Geometry and Spatial Form." But I am not disputing Friedman's point that the possibility of the abstraction that Kant describes in this passage presupposes the pure form of spatial intuition, only Friedman's supposition that (repeated) construction of figures in pure intuition presents us with an actual singular space.

[15] See for example Reflection 6315, 18:618–21, at 18: 619; *Notes and Fragments*, pp. 361–3, at p. 361.

> on the original representation of a single, infinite *subjectively given* space. It is entirely consistent with this that the geometrical and objectively given space is always *finite*; for it is only given insofar as it is *made*. But that the metaphysical, i.e., original but merely subjectively given space, which cannot be brought under any concept that would be capable of construction (because there are not many of them), but yet contains the ground of the construction of all possible geometrical concepts, is *infinite*, says only that it consists in the pure form of the sensible mode of representation of the subject as intuition *a priori*, consequently in this, as a singular [*einzeln*] representation, the possibility of all spaces, that extend to the infinite, is *given*. (20:420–1)

This clearly says that metaphysical space is just the pure form of the intuition of space. It is not an object at all, but a form. As a form, it is given, to be sure. But as a mere form, it also has no determinate magnitude. On the basis of this form, we can construct geometrical representations of space – lines, plane figures, etc. – of any size, and for any figure of any size we can construct another of greater size. So this pure form allows the indefinite expansion of any kind of spatial figure, although we cannot literally construct or "make" a figure of infinite size because we cannot complete an infinite synthesis. So we are not given infinite figures, and we are not given the infinitude of space itself – we have to infer it from the character of the form of spatial representation as a container, which is what we are immediately given.

11.3 The Unity of Space and Time in the Transcendental Deduction

I can now make my point about the Transcendental Deduction more briefly. As is well known, Kant entirely rewrote the central section of the Deduction for the second edition of the *Critique of Pure Reason*. He divided it into thirteen numbered sections (§§15–27), and marked its division into two halves by stating in §21 that the proposition proven in §§15–20, namely that "A manifold that is contained in an intuition that I call mine is represented as belonging to the **necessary** unity of self-consciousness through the synthesis of the understanding, and this takes place by means of the category," represents just "the beginning of a **deduction** of the pure concepts of the understanding." Kant explains that in the argument leading up to that proposition he has "abstract[ed] from the way in which the manifold for an empirical intuition is given, in order to attend only to the unity that is added to the intuition through the understanding by means of the category," but that in what follows the deduction will be completed by "show[ing] from the way in which the empirical intuition is given in sensibility that its unity can be none other than the one the category prescribes to the manifold of a given intuition in general" (B 144). In other words, the Transcendental Deduction will be completed by appealing to the specifically spatio-temporal character of human sensible intuition, and by that means the categories will be shown to apply necessarily to all human intuitions and to objects as presented by them: the initial doubt – a Cartesian doubt in the sense of being staged only to be resolved (A 89/B 122) – that because the categories are not immediately given in or with sensible intuition in the way that spatio-temporal form is they might not apply to sensible intuition at all will thereby be allayed. Much of what follows §21, however, focuses on the *limits* on the use of the categories that follow from their necessary application to sensible intuition, namely that they yield knowledge only of appearances, not things in themselves; Kant's positive argument that the categories *do* necessarily apply to all sensible intuition seems to be presented only in §26.

Since Dieter Henrich published his article on "The Proof-Structure of the Transcendental Deduction" in 1969,[16] there has been much discussion of what the relation between the two parts of the deduction thus marked out is supposed to be. Henrich's position was that the first stage of the proof left the *scope* of the application of the categories unspecified, arguing only that any manifold of intuition that is unified and ascribed to a self-conscious self must be unified and ascribed to the self through the use of the categories, while the second stage lifts that "restriction" by arguing that *all* of one's spatio-temporal manifold of intuition must be able to be unified into the representation of a single spatio-temporal world and ascribed to one's single self. However, this approach is open to the objection that Kant begins the first stage of the argument with the premise that "The **I think** must **be able** to accompany *all* my representations" (§16, B 131; italics added), so there is no restriction on the scope of the application of the categories in the first stage of the argument even though there is also no explicit reference to the specifically spatio-temporal form of the intuitions to which the categories are applied in it as the necessary condition of being able to accompany all my representations with the "I think." For this reason, other interpretations of the two-stage structure of the deduction have sought to find different explanations of the significance of the introduction of the specifically spatio-temporal character of our sensible intuition in the second stage.

In his recent book on the Transcendental Deduction, Henry Allison has largely followed this alternative to Henrich's strategy.[17] Much of his interpretation of the second half of the revised deduction is that the introduction of spatio-temporal form into the argument allows for the explanation of *how* the categories apply to intuition, namely through what Kant calls the "figurative synthesis" exercised by the "productive imagination" as explained in §24. (The second stage of the second-edition deduction thus plays the role of what in the first-edition Preface Kant had called the "subjective" side of the deduction, which "deals with the pure understanding itself, concerning its possibility and the powers of cognition on which it itself rests," a "matter of great importance in respect of" Kant's "chief end" but not "essential" to it; A xvi–xvii). But when it comes to §26, Allison's interpretation seems to take a Henrichian turn, for what he suggests is that in this section Kant argues that all synthesis of empirical intuitions, beginning with the "empirical synthesis of apprehension," must "conform to the conditions imposed by the transcendental synthesis"

[16] Dieter Henrich, "The Proof-Structure of the Transcendental Deduction," *Review of Metaphysics* 22 (1969): 640–59.

[17] Henry E. Allison, *Kant's Transcendental Deduction: An Analytical-Historical Commentary* (Oxford: Oxford University Press, 2015).

of the pure manifold of intuition into a "single time and space" governed by the categories, thus "the empirical synthesis" (of the manifold of empirical intuition) "must likewise conform to the rules of unification prescribed by the categories."[18] Thus the categories must apply to all empirical intuitions because all empirical intuitions must be synthesized within the framework of a single space and a single time, and the categories are the necessary conditions of the synthesis of the pure manifolds of space and time into such single frameworks. The point of this argument appears to be precisely to prove that the categories necessarily apply to *all* of anyone's possible empirical intuitions, thus apparently to lift a restriction presumably contained in the earlier stage of the argument. That is why I claim that Allison falls back into Henrich's approach in this last part of his interpretation.

But it is on this last point that I want to press, not so much to dispute the accuracy of his interpretation of Kant but rather to raise doubt about the key assumption of what may well have been Kant's own argument. With close attention to the first three paragraphs of §26 and the footnote appended to the last of these at B 160–1, Allison interprets Kant's argument as turning on the claim that synthesis in accordance with the categories is necessary to get from time as a mere *form of intuition* to time as a pure *formal intuition*, that is, to an *a priori* representation of the unity of time itself, as contrasted to the unity of all the states of affairs in time, which would be the subject of the empirical synthesis that is to be governed by the *a priori* synthesis of the unity of time itself. (Allison focuses on the unity of time because of the supposition that time is the form of all intuition, while space is the form only of outer intuition, although as we will see Kant's own argumentation in §26 itself focuses on the case of space.) Borrowing the terminology introduced by Onof and Schulting,[19] Allison puts the point by saying that the categories are required to establish the *unity* rather than mere *unicity* of time (or space) itself: "unicity" is "a mereological property that belongs to space and time as *given*," namely that any part of space or time is immediately given as part of some larger space or time, which is something that we immediately intuit, in pure intuition, and requires no synthesis of any kind, while the "unity" of space and time pertain to "the *representation* of their unity by the cognizing subject" and must be "a product of a transcendental synthesis,"[20] executed in accordance with the categories and in turn governing the synthesis of the representation of the empirical world in (space and) time, which is therefore also subject to the categories – which is, of course, what is to be proven by the completed deduction. As Allison sums up, "The point is that the conditions of the representation of space or time are also conditions of the apprehension of anything in space or time" because "the empirical synthesis of apprehension, which is concerned with the unification of empirical manifolds given in space and time, is subject to the conditions of the transcendental synthesis of the imagination, which is charged with the unification of the pure manifold of space and time."[21]

[18] Allison, *Transcendental Deduction*, p. 406.

[19] Onof and Schulting, "Space as Form of Intuition and as Formal Intuition," especially pp. 12–16.

[20] Allison, *Transcendental Deduction*, p. 412.

[21] Allison, *Transcendental Deduction*, p. 415.

The problem with this argument lies in the move from the "unicity" of space and time, which needs no transcendental synthesis, because it is what is immediately given in pure intuition, namely that any limited region of space or time must be represented as part of a larger one because it is indeed represented at all only by being delimited within such a larger space or time, to the unity of the representation of space and time as unities or wholes, which allegedly does require a synthesis in accordance with the categories which are in turn imposed upon all empirical synthesis.[22] I reiterate that my worry is not with the accuracy of Allison's interpretation, which may well be the natural reading of Kant's claim in the well-known footnote to §26, although Kant makes it with regard to space rather than time, that "Space, represented as **object** (as is really required in geometry), contains more than the mere form of intuition, namely the **comprehension** of the manifold given in accordance with the form of sensibility…so that the **form of intuition** merely gives the manifold, but the **formal intuition** gives unity of the representation," which unity "presupposes a synthesis, which does not belong to the senses but through which all concepts of space and time first become possible" (B 160-1n.). The problem is with Kant's claim itself. The problem is with the supposition that there *is* a representation of the unity of (space and) time as a *synthetic* unity or totality of all its parts that does intervene between the mere form of intuition with its "unicity" and the synthetic unity of the representation of the empirical world in space and time. More generally, the problem is that on Kant's own account, as we previously saw, there is no such thing as the pure mathematical representation of completely determinate objects: mathematics, including therein the mathematical representation of space and time, never represents more than the form of possible objects. Actual objects are represented only empirically, and the determinate scope of any object is something that can be represented only empirically. Mathematics does not actually represent the unity of space or time; *a fortiori* there is no transcendental synthesis of the whole of space and time, guided by the categories, to precede the synthesis of empirical intuitions and impose the categories on the latter.

To my previous argument that geometrical constructions are not actually representations of determinate objects, thus cannot very well be used as a model for the representation of space as a determinate (even if unlimited or infinite) whole, I can now add that there is no manifold of parts to be synthesized that is *given* in the pure intuition of space as there is in temporally extended empirical intuition. There is simply the fact that I can imagine some lines and their intersections, or a smaller figure contained in a larger one, or a smaller space contained in a larger one. But I

[22] It will be recalled that Onof and Schulting had included infinitude among the marks of unicity; they assumed that infinitude is given, while I argued that it can only ever be represented as the product of an incompletable synthesis, thus as an idea of reason. Allison does not employ the assumption of the infinitude of space or time; his argument is rather that an *a priori* synthesis of the *unity* of space and time precedes empirical synthesis of the world of objects in space and time, and that the necessity of the categories for the latter is guaranteed by their role in the former. Thus I might say that my real target here is my old enemy, the idea of an *a priori* synthesis that precedes empirical synthesis; see my "Kant on Apperception and *A Priori* Synthesis," *American Philosophical Quarterly* 17 (1980): 205–12.

am not *given* determinate parts of space in pure intuition, thus there are no parts of pure space or pure manifold of space waiting to be unified. Similarly, in the case of time, we can immediately see in the exercise of pure intuition that any period of time can be contained in a longer period of time, and that any period of time can be subdivided into shorter periods of time. But there are no parts of time *given* in pure intuition to be synthesized into a whole of time in a transcendental synthesis that could precede empirical synthesis. Kant himself makes this point when he states that "Every intuition contains a manifold in itself, which however would not be represented as such if the mind did not distinguish the time in the succession of impressions on one another" (A 99): the "parts" of time are not given in pure intuition to be synthesized in a pure synthesis, but are introduced into the form of time, and indeed only in experience. We have no actual periods of time to synthesize into a whole until we demarcate some empirically, for example as the periods between successive noons or full moons or oscillations of some crystal. Only once we have empirically represented particular periods of time can we represent the whole of time, or move in that direction (since we cannot complete an infinite synthesis). In pure intuition there is neither a whole of time to be divided into parts nor parts of time to be synthesized into a whole, only the recognition that the form of time is to be always further extendable or further divisible.

These are general grounds for doubting the very idea that there is a pure synthesis of the wholes of space and time which in any way conditions empirical synthesis; because pure intuition gives only possible forms of spatio-temporal objects, it is only in empirical synthesis that actual spatio-temporal objects, including space and time themselves as wholes, are synthesized. But even leaving this general point aside, the idea of a category-dependent synthesis of pure space and pure time as wholes sufficient for the purposes of the deduction is problematic. On which categories, exactly, would such a synthesis depend? I argued earlier that an infinite given magnitude of space or time should not be regarded as immediately intuited because even if we immediately intuit that any space or time is surrounded by a larger one, we have to reiterate that insight in order to arrive at the idea of infinite space or time. The act of iteration can be regarded as a synthesis of the understanding – that would be a way of putting my original objection to Kant's claim – and perhaps we could even argue that it uses *some* of Kant's categories, namely the categories of quantity, that is, the concepts of units, pluralities (partial collections of units), and totality (A80/B 106). Perhaps the synthesis here also uses one of the modal categories – namely, possibility – although given Kant's subjective interpretation of modality, that category should properly be applied to the synthesis itself, not used in the synthesis or applied to its object. Maybe we could say that the understanding's synthesis of the idea of a larger or infinite space or time uses some of the relational categories: perhaps we conceptualize space or time as a substance of which the parts introduced by delimitation are accidents, maybe we conceive of the parts of space and time as standing in relations of ground and consequence or community.[23] But

[23] On the use of the categories for synthesis of the unity of space, see Onof and Schulting, pp. 42–4. The argument for the use of the full set of the categories depends on their role in the construction of geometrical figures, and not just on their use for the supposed synthesis of the unity of pure space itself.

we certainly cannot conceive of parts of pure space or time as such as subsumed under the *schematized* categories, such as that of substance enduring *in* time or cause-and-effect rather than ground-and-consequence, and *those* are the ones that Kant is eager to prove necessarily apply to all *empirical* intuitions. That could not be proved even by proving that (some) pure categories are necessarily involved in the synthesis of pure space and time as wholes, were there any such thing.

The point is evident in Kant's own illustration ("e.g.") of the claim that "synthetic unity can be none other than that of the combination of a given **intuition in general** in an original consciousness, in agreement with the categories, only applied to our **sensible intuition**" (§26, B 161) – a proposition that should not be read to assert that there is a pure synthesis of the unity of space or time that underlies the synthesis of empirical intuitions in space and time, but rather that the conditions that can be abstractly described as those of the synthesis of a manifold of intuition in general are *in our case* applied in the synthesis of a spatio-temporal manifold of empirical intuitions. Kant makes this clear when he writes:

> If (in another example) I perceive the freezing of water, I apprehend two states (of fluidity and solidity) as ones standing in a relation of time to each other. But in time, on which I ground the appearance as **inner intuition**, I represent necessary synthetic **unity** of the manifold, without which that relation could not be **determinately** given in an intuition (with regard to the temporal sequence). But now this synthetic unity, as the *a priori* condition under which I combine the manifold of an **intuition in general**, if I abstract from the constant form of **my** inner intuition, time, is the category of **cause**, through which, if I apply it to my sensibility, **I determine everything that happens in time in general as far as its relation is concerned.** (§26, B 162–3)

Even if the representation of the whole of the time in which the states of fluidity and solidity were to require some kind of pure synthesis, all that the representation of time as such would tell us is that if those two states are states of one object then one of them must precede and one succeed the other. It does not tell us which. Only the application of the schematized category of causality, in the form of a particular causal law, in this case the law that a drop of temperature to zero Celsius will cause water to transform from its liquid to its solid state, will tell us that. And what this means is that the applicability of the categories to our empirical intuitions cannot be proven from their role in the synthesis of the supposed pure manifolds of space and time into the supposed wholes of pure space and time, but only from their role in enabling the synthesis of *empirical* intuitions into the representation of a determinately ordered world of objects and their states in actual, so to speak *empirical* space and time. Thus, when Onof and Schulting state that "our objective experience is *one world of outer sense*. That is, objectivity requires the singularity of the whole domain of outer sense,"[24] that is correct, but it does not follow from anything about the pure form of space as given in pure intuition. It is required by the transcendental unity of apperception, but has to be established on the basis of the Analogies of Experience.

[24] Onof and Schulting, "Space as Pure form and as formal Inution," p. 21.

This example obviously anticipates Kant's argument in the second Analogy of Experience. Although he does not anticipate it here, the third Analogy is also relevant to the present argument. Kant's claim there is that (states of) different objects can be cognized as simultaneously located at different regions of space only if they physically interact with each other. The details of the argument are not important for us here; what is important is that while the pure form of space may *allow* for the location of different objects at different locations simultaneously, the determination that different objects are at different locations at the same time is not made on the basis of any category-dependent synthesis of pure space and time as wholes, but on the basis of the application of the category of interaction, through particular laws of interaction (say, laws specifying specific relations of attraction and repulsion), to empirical objects in real, again empirical space and time. Even if there were such a thing as a category-dependent synthesis of pure space and time, the necessary applicability of the category of interaction to empirical intuitions would not follow from that, but from the conditions of determining the relative simultaneous locations of empirical objects in real space and time.

All of this can be illustrated in a response to a once-famous thought experiment by Anthony Quinton.[25] He imagined a subject who experienced what seemed to be one coherent spatio-temporal world by day, day after day, when he was awake, and then another one at night, night after night, when he was dreaming, and then asked how the subject could determine the relation between those two represented worlds. The answer is that the subject could not determine there to be any determinate relation between those worlds unless he could establish determinate relations of causation and interaction between objects and events in them, thus, in the terms I have been using here, that he could not determine the relation between these two worlds on the basis of the pure form of intuition alone. Even though the spatio-temporal forms of human sensible intuition might determine that *within* each series of experience the subject envisioned what he experienced as the experience of a single spatio-temporal whole, that would not suffice to determine that he really did experience one spatio-temporal whole. Because any time we represent objects in space and time we represent them as contained in a larger, unitary space and time, we might say that the subject has no choice but to *try* to represent his two different series of experiences as parts of one larger space and time, but if he can find no determinate relations of causation and interaction between the two series he cannot actually *succeed* in doing so.[26] And this is just to say that the mere *forms* of spatial

---

[25] Anthony Quinton, "Spaces and Times," *Philosophy* 37 (1962): 130–47.

[26] It might be objected that Quinton's thought-experiment depends upon the assumption that the subject must at least be able to represent himself as a single subject in both of his sequences of his experiences, thus at least remember having the night-time experiences while having the day-time experience, in order to even ask whether he is in fact the same subject through both series of experiences. But if this just takes the veracity of memory for granted, when in fact memory cannot function without a framework of objective knowledge to support it, all the better for my argument: the subject could not recognize himself as one without the representation of a single spatio-temporal world, but a single world of empirical objects in space and time, not some single pure intuition of space and time, which as I have been arguing is only a representation of a single form but not a single object. Thanks to Rosalind Chaplin for this objection.

and temporal representation as the representation of smaller regions within larger ones is *not* by itself sufficient to constitute the experience of a single space and time. That depends on the empirical synthesis of empirical intuitions in accordance with the schematized categories, and is not given by the pure form of intuition or any synthesis of that alone.

That there is a whole of pure space and time to be synthesized in a transcendental synthesis that underlies the empirical synthesis of a world of appearances from empirical intuitions and guarantees the applicability of the categories to the latter is thus a myth. The only synthesis clearly involving all the categories is the empirical synthesis of the empirical world, in accordance with the transcendental conditions represented by the pure forms of intuition, as forms of empirical intuitions, and the pure concepts of the understanding, as forms of empirical concepts. What does this conclusion tell us about the relation of the two parts of the second-edition Transcendental Deduction? The first part proposes a guarantee of the possibility of the thoroughgoing, schematized-category-guided empirical synthesis of empirical intuitions in the form of the transcendental unity of apperception, the possibility of attaching the "I think" to each and all of one's empirical intuitions. For reasons we have not explored here, this is presumed to require their synthesis in accordance with the categories. This already settles any possible question about the scope of the application of the categories. The second half of the deduction can only tell us more about how the categories apply to our empirical intuitions, as indeed Allison originally proposed. Its argument begins in §22 with the claim that pure mathematics does not give us actual cognition except insofar as it is applied to empirical objects. It concludes with the inference that the categories are to be applied to intuitions with spatio-temporal form, but this must be understood as an argument that the synthesis of our empirical intuitions into a determinate spatio-temporal world, required by the transcendental unity of apperception, depends upon the categories, rather than an argument that there is a synthesis of pure space and time as synthetic unities or totalities, since there is no good ground for positing such a synthesis. In a footnote, Kant takes a stab at illustrating the role of the categories in the synthesis of spatio-temporal representations with the case of geometrical figures, but in this footnote he neither explicitly states that there is a synthesis of pure space and time themselves as synthetic unities nor explains how the synthesis of geometrical figures involves all the categories. It seems a sound hermeneutical strategy to place more weight on the examples of spatio-temporal synthesis that Kant includes in the main text of the section rather than in a footnote, and these examples suggest that the synthesis needed to redeem the promise of the transcendental unity of apperception is the synthesis of empirical intuitions into determinate representations of empirical objects such as houses and freezing water that can in turn be synthesized into the determinate representation of an empirical world, a spatio-temporal whole of determinately related empirical objects as opposed to a domain of indeterminate and unrelated mere forms of spatio-temporal objects. Of course, the details of the synthesis of the empirical world are more fully spelled out only in the Second Book of the Transcendental Analytic, the Analytic of Principles, particularly in its Second Chapter, on the System of All Pure Principles of Pure Understanding. The second half of the second-edition deduction is thus itself just a promissory note, or, if this sounds better, the transition from Kant's abstract or transcendental conception of the unity of apperception to his exposition of the conditions of the possibility of empirical self-consciousness in a world of empirical objects in real space and time.[27]

Abbreviations

A/B Kant, Immanuel. *Critique of Pure Reason*, edited and translated by Paul Guyer and Allen W. Wood. Cambridge: Cambridge University Press, 1998. A refers to the 1781 edition and B to the 1787 edition.

ID "On the Forms and Principles of the Sensible and Intelligible Worlds", *Theoretical Philosophy 1755–1770*, translated and edited by David Walford in collaboration with Ralf Meerbote. Cambridge: Cambridge University Press, 1992.

References

Allison H. E. (2004). *Kant's transcendental idealism: An interpretation and defense* (Revised ed.). New Haven: Yale University Press.

Allison, H. E. (2015). *Kant's transcendental deduction: An analytical-historical commentary*. Oxford: Oxford University Press.

Dieter, H. (1969). The proof-structure of the transcendental deduction. *Review of Metaphysics, 22*, 640–659.

Friedman, M. (2012). Kant on geometry and spatial intuition. *Synthese, 186*, 231–255.

Guyer, P. (1980). Kant on apperception and *a priori* synthesis. *American Philosophical Quarterly, 17*, 205–212.

Guyer, P. (2001). Space, time, and the categories: Kant's project in the transcendental deduction. In R. Schumacher & O. R. Scholz (Eds.), *Idealismus als theorie der repräsentations?* (pp. 313–338). Paderborn: mentis Verlag.

Kant, I. (2005). *Notes and fragments*, edited by Paul Guyer, trans. by Bowman C. Guyer P., & Ruascher F. Cambridge: Cambridge University Press.

McLear, C. (2015). Two kinds of unity in the critique of pure reason. *Journal of the History of Philosophy, 53*, 79–110.

Onof, C., & Schulting, D. (2015). Space as form of intuition and as formal intuition: On the note to B 160 in Kant's critique of pure reason. *Philosophical Review, 124*, 1–58.

Quinton, A. (1962). Spaces and times. *Philosophy, 37*, 130–147.

[27]On this point, see also my paper "Space, Time, and the Categories: Kant's Project in the Transcendental Deduction," in Ralph Schumacher and Oliver R. Scholz, eds., *Idealismus als Theorie der Repräsentation?* (Paderborn: Mentis Verlag, 2001), pp. 313–38.

Author Index

O. Nachtomy, R. Winegar (eds.), *Infinity in Early Modern Philosophy*, The New Synthese Historical Library 76,
https://doi.org/10.1007/978-3-319-94556-9

Subject Index

O. Nachtomy, R. Winegar (eds.), *Infinity in Early Modern Philosophy*, The New Synthese Historical Library 76,
https://doi.org/10.1007/978-3-319-94556-9

Kolibri 360 GmbH
Ferdinand-Jühlke-Straße 7
99095 Erfurt, Deutschland
produktsicherheit@kolibri360.de